Notes on Medical Bacteriology

J. Douglas Sleigh
Reader in Bacteriology

Morag C. Timbury
Professor and William Teacher Lecturer in Bacteriology

The Department of Bacteriology,
University of Glasgow,
Royal Infirmary, Glasgow

FOREWORD BY

Sir James W. Howie
Formerly Director, Public Health Laboratory Service

SECOND EDITION

CHURCHILL LIVINGSTONE
EDINBURGH LONDON MELBOURNE AND NEW YORK 1986

CHURCHILL LIVINGSTONE
Medical Division of Longman Group UK Limited

Distributed in the United States of America by Churchill
Livingstone Inc., 1560 Broadway, New York, N.Y.
10036, and by associated companies, branches and
representatives throughout the world.

First published 1981
Second edition 1986
Reprinted 1986

British Library Cataloguing in Publication Data
Sleigh, J. Douglas
    Notes on medical bacteriology. — 2nd ed.
    (Churchill Livingstone medical text)
    1. Bacteriology, Medical
    I. Title II. Timbury, Morag C.
    616'.014   QR46

Library of Congress Cataloging in Publication Data
Main entry under title:

Notes on medical bacteriology.

    Based on the lecture course given to medical
students at the University of Glasgow.
    Bibliography: p.
    Includes index.
    1. Bacteriology, Medical.   2. Bacterial diseases.
I. Sleigh, J. Douglas.   II. Timbury, Morag Crichton.
[DNLM: 1.   Bacteriology—handbooks. QW 39 N911]
QR46.N68   1986   616'.014   85-19473

Produced by Longman Group (FE) Ltd
Printed in Hong Kong

ISBN 0-443-03327-7

Notes on Medical Bacteriology

LORD LISTER
Professor of Surgery in Glasgow Royal Infirmary
from 1861 to 1869

Foreword

These admirable notes represent an entirely successful effort to present the main facts of medical microbiology in the fewest possible words. They will surely come as a boon and a blessing to all students who require an introduction to the subject which picks out the information that matters most from the now embarrassingly large volume of good published work. This mass of riches makes the work of authors of supposedly elementary textbooks very difficult indeed. The essence of art is selection, but the selection must be properly balanced, and therein lies the difficulty. In teaching medical microbiology to medical undergraduates, to young medical graduates who are beginning to specialise in the subject, and to science graduates and scientific officers who are to work in medical laboratories, it is important to give as an introduction enough about the biology of the microbes to make clear how they may be identified so that the correct specimens are sent for examination, the correct methods used to that end, and the correct interpretations put upon the laboratory findings. Matters of detail and method soon become second nature to experienced microbiologists in their day-to-day work; but the new student must not be put off the subject by thinking that he must memorise from the beginning all the details in the cookery book of the microbiologists' kitchen. What these notes do so well is to set out the necessary basic facts and principles clearly and briefly for noting and easy reference, and then to concentrate on the exciting facts about where microbes live, how they get around, how they may be contained, what diseases they produce, and how these diseases may be prevented or treated. The concerns both of individual patients and of the community are kept in proportion. Immunisation, hygiene, epidemiology, and antimicrobial therapy are all competently dealt with alongside the facts about the microbes concerned. The result is a miniature that is also a masterpiece. I have pleasure in writing this foreword and in wishing the book the success I know it deserves.

J. W. H.

Preface

This book is intended for medical students studying for the Professional Examination in Microbiology and is based on the course given to those in the University of Glasgow who attend our classes in the Royal Infirmary. It aims to give a concise account of medical bacteriology. Systematic bacteriology is dealt with briefly and we have tried to emphasise the clinical aspects of the subject which seem to us of most importance in present-day medical practice. It should be supplemented by reading from a larger textbook and some recommended books for further reading are listed on page 392.

For this edition, the book has had considerable revision and two sections on medical parasitology and mycology respectively have been added.

We are again grateful to colleagues who have helped us with discussion and advice, notably Dr D.R. Baird, Miss M. Bruce, Dr J.R. Donaldson, Dr R.J. Fallon, Dr K. Hare, Mrs J. McCabe and Mr I. Marshall.

We thank Professor J.G. Collee and Drs A.K.R. Chaudhuri, D.H.M. Kennedy, M. Laidlaw and W.C. Love, for supplying us with colour photographs. The drawing in Plate 1 is based on an original idea of Dr D.J. Platt.

We thank Mrs M. Meinertz and Miss G. Wink for the typing of the manuscript.

Both of us owe a particular debt to Sir James Howie not only for writing the foreword to the book but for our early years of training in his department. He inspired a whole generation of medical microbiologists.

GLASGOW,   J. Douglas Sleigh
1986   Morag C. Timbury

Acknowledgements

We wish to acknowledge the help we have received in the preparation of the following chapters:

Growth and nutrition of bacteria
Host-parasite relationship
Dr C.G. Gemmell

Medical Mycology
Dr G.R. Jones

Infections in general practice
Dr C. Langan

Infections of the eye
Professor W.R. Lee

Bacteria: organisation, structure, taxonomy
Dr D.J. Platt

Parasitology
Dr H. Williams

We are grateful to the following for the illustrations for this edition:

Diagrams and drawings
Mr J.G. Ramsden

Photographs
Mr W. Patterson

Photomicrographs
Mr J. Winning

Contents

Bacterial biology

1

Introduction

Medical students need to learn bacteriology in order to diagnose and treat bacterial infections successfully.

Bacterial disease is still widespread and common but its spectrum is changing because once familiar diseases are now rare and new infections are being recognised. Increasingly, the work of bacteriology laboratories is concerned with infection in patients in general hospitals—as distinct from fever hospitals—and in general practice.

The following are among the most important aspects of bacteriology which doctors must know in order to deal with infection.

Pathogenesis. The ways in which bacteria produce disease in the human body—essential information for diagnosis and treatment.

Diagnosis. Laboratory investigation depends on taking correct specimens and being able to assess the results obtained from the laboratory.

Treatment. Bacterial disease is one of the few conditions in medicine for which specific and highly effective therapy is available.

Epidemiology. The spread, distribution and prevalence of infection in the community.

Prevention. Many bacterial diseases have been virtually eradicated by immunization, public health measures and improved living standards.

HISTORY

Contagion. Since biblical times it has been known that some diseases spread from person to person.

The following are some of the pioneers responsible for the science of bacteriology as it is today.

Antony van Leeuwenhoek: a Dutch draper who made a microscope and in 1675 observed 'animalcules' in samples of water, soil and human material.

Louis Pasteur, the founder of modern microbiology: over a long

3

period of brilliant and active research from 1860 to 1890, he developed methods of culture and showed that microorganisms cause disease. He also established the principles of immunisation.

Joseph Lister was Professor of Surgery in Glasgow Royal Infirmary. He applied Pasteur's observations to the prevention of wound sepsis — then almost an inevitable and often fatal complication of surgery. He discovered in 1867 an antiseptic technique to kill bacteria in wounds and in the air with carbolic acid. This revolutionised surgery.

Robert Koch was a German general practitioner who discovered the bacterial causes of many diseases—including tuberculosis in 1882. He introduced agar as a setting agent for bacteriological media although the discovery is attributed to Frau Hesse from observations made in her kitchen. Koch defined the criteria for attributing an organism as the cause of a specific disease. These are the famous **Koch's postulates** and are as important today as when he propounded them:

1. The organism is found in all cases of the disease and its distribution in the body corresponds to that of the lesions observed.
2. The organism should be cultured outside the body in pure culture for several generations.
3. The organism should reproduce the disease in other susceptible animals.
   Nowadays a fourth postulate would be added:
4. Antibody to the organism should develop during the course of the disease.

Note. Many infectious diseases of which the cause is clearly identified do not fulfil the third and even occasionally the second of Koch's postulates.

Immunisation

The first successful immunisation was the demonstration by *Edward Jenner* in 1796 that a related but mild virus disease—cowpox—gave protection against subsequent attack by smallpox. Later, Pasteur's observations led to the development of the vaccines now widely and successfully used in medicine against diseases, e.g. diphtheria, tetanus, poliomyelitis, etc.

Antibiotics

The discovery of penicillin in 1929 by *Alexander Fleming*—a Scot from Ayrshire—ushered in the antibiotic era. Generally derived from soil microorganisms, antibiotics kill or inhibit a wide variety of bacteria without harming their human host.

Public health

The development of a safe water supply, disposal of sewage, good housing and nutrition have also been major factors in the decline of epidemic infectious disease.

2

Bacteria: organisation, structure, taxonomy

Bacteria are a heterogeneous group of unicellular organisms: their cellular organisation is described as *prokaryotic* (i.e. having a primitive nucleus) and differs from that of *eukaryotes* (plants, animals). Some of the main differences are listed in Table 2.1.

Table 2.1 Differences between prokaryotic and eukaryotic cells

Property	Prokaryotic cells	Eukaryotic cells
Chromosome number	One	Multiple
Nuclear membrane	Absent	Present
Mitochondria	Absent	Present
Sterols	Absent	Present
Ribosomes	70s	80s

Genome. The most fundamental difference between bacteria and eukaryotes is that the bacterial chromosome, or genome, is a single circular molecule of double-stranded DNA; there is no nuclear membrane. Bacteria may from time to time harbour other smaller circular DNA molecules or plasmids which code for certain non-essential functions.

STRUCTURE

Bacteria have a rigid wall which determines their shape

Shape. Bacteria may be:
1. Spherical—cocci
2. Cylindrical—bacilli
3. Helical—spirochaetes

Arrangement depends on the plane of successive cell divisions: e.g. chains—streptococci; clusters—staphylococci; diplococci—pneumococci; angled pairs of palisades—corynebacteria.

Gram's stain divides bacteria into Gram-positive or Gram-negative,

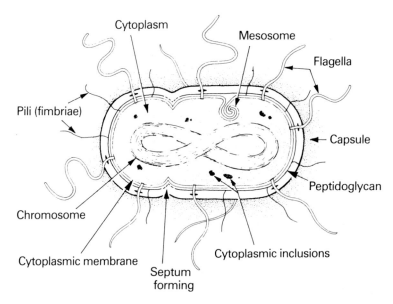

Fig. 2.1 Diagram of a typical but composite bacterial cell.

an important step in classification. The Gram-staining reaction depends on the structure of the cell wall.

A diagram of a typical but composite bacterium is shown in Figure 2.1 and in Plate 1. Bacteria are cells with a rigid cell wall which surrounds the protoplast; this consists of a cytoplasmic membrane enclosing internal components and structures such as ribosomes and the bacterial chromosome.

External structures

External structures which protrude from the cell into the environment are present in many bacteria. These structures are:

1. *Flagella:* helical filaments which produce motility by rotation; composed of protein sub-units — 'flagellin' (Fig.2.2).
2. *Pili:* finer shorter filaments extruding from the cytoplasmic membrane; also protein (pilin), they are responsible for adhesion (common pili) and probably for conjugation when genes are transferred from one bacterium to another (sex pili).
3. *Capsules:* amorphous material which surrounds many bacterial species as their outermost layer; usually polysaccharide, occasionally protein; often inhibit phagocytosis and so their presence correlates with virulence in certain bacteria.

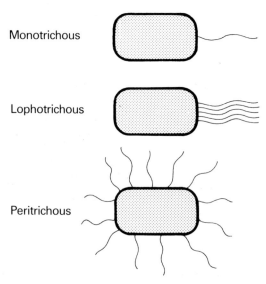

Monotrichous

Lophotrichous

Peritrichous

Fig. 2.2   Distribution of flagella on bacteria.

Gram-positive cell wall

In addition to conferring rigidity upon bacteria the cell wall also protects against osmotic damage. It is porous and permeable to substances of low molecular weight.

Structure of the cell wall differs in Gram-positive and Gram-negative bacteria; this is illustrated in Figure 2.3.

Chemically the rigid part of the cell wall is peptidoglycan: this is a mucopeptide composed of alternating strands of N-acetyl muramic acid and N-acetyl glucosamine cross-linked with peptide subunits (Fig. 2.4).

Gram-negative cell wall differs from that of the Gram-positive bacteria by the presence of an *outer membrane* which contains specific proteins (outer membrane proteins). These form 'porins' through which hydrophilic molecules are transported. Other proteins are receptor sites for phages and bacteriocins. The lipopolysaccharide O antigens are embedded in the outer membrane.

Gram-positive cell wall. The peptidoglycan layer is much thicker than in Gram-negative bacteria. There is no periplasmic space and the peptidoglycan is closely associated with the cytoplasmic membrane. Figure 2.5 is a diagrammatic representation of a Gram-positive cell wall.

Teichoic or teichuronic acids are part of the cell wall of Gram-positive

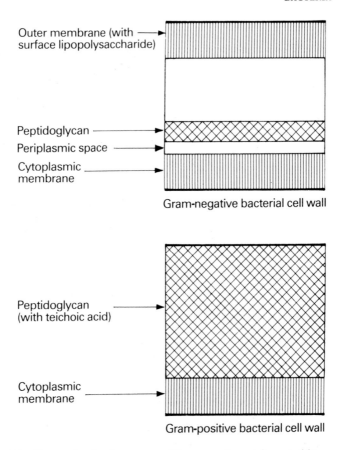

Gram-negative bacterial cell wall

Gram-positive bacterial cell wall

Fig. 2.3 Diagram showing the structure of Gram-negative and Gram-positive bacterial cell walls.

bacteria: they maintain the level of divalent cations outside the cytoplasmic membrane.

Other components which may be present in the cell wall are antigens such as the polysaccharide (Lancefield) and protein (Griffith) antigens of streptococci.

Bacteria with defective cell walls

Bacteria develop and can survive with defective cell walls: these can be induced by growth in the presence of antibiotics and a hyperosmotic environment to prevent lysis.

Bacteria without cell walls are of four types:

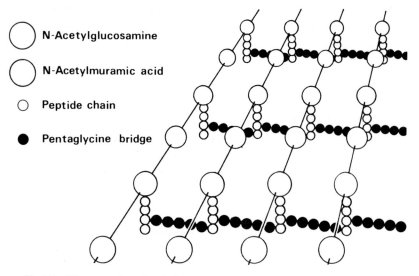

- ○ N-Acetylglucosamine
- ○ N-Acetylmuramic acid
- ○ Peptide chain
- ● Pentaglycine bridge

Fig. 2.4 Diagram to show chemical structure of cross linking in peptidoglycan component of cell walls. From Sharon N The Bacterial Cell Wall. Copyright (C) 1969 by Scientific American Inc. All rights reserved.

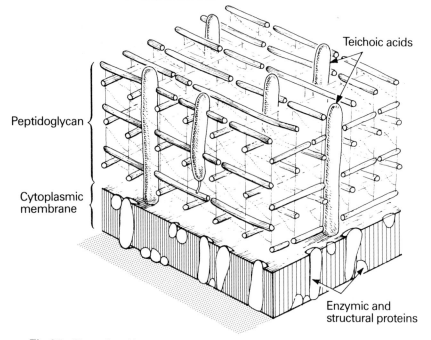

Teichoic acids

Peptidoglycan

Cytoplasmic membrane

Enzymic and structural proteins

Fig. 2.5 Three-dimensional representation of Gram-positive bacterial cell wall.

1. *Mycoplasma:* an independent bacterial genus of naturally occurring bacteria which lack cell walls; also stable and do not require hypertonic conditions for maintenance.

2. *L-forms:* cell wall-deficient forms of bacteria usually produced in the laboratory but sometimes spontaneously formed in the body of patients treated with penicillin; more stable than protoplasts or spheroplasts, they can replicate on ordinary media.

3. *Spheroplasts:* derived from Gram-negative bacteria; retain some residual but non-functional cell wall material; osmotically fragile; produced by growth with penicillin and must be maintained in hypertonic medium.

4. *Protoplasts:* derived from Gram-positive bacteria and totally lacking cell walls; unstable and osmotically fragile; produced artificially by lysozyme and hypertonic medium: require hypertonic conditions for maintenance.

Cytoplasmic membrane

A 'unit membrane' i.e. a double-layered structure composed of lipid and protein which acts as a semipermeable membrane through which there is uptake of nutrients by passive diffusion. It is also the site of numerous enzymes involved in the active transport of nutrients and various other cell metabolic processes.

Mesosomes

Convoluted invaginations of cytoplasmic membrane often at sites of septum formation: involved in DNA segregation during cell division and respiratory enzyme activity.

Nuclear material

The single circular chromosome which is the bacterial genome or DNA undergoes semiconservative replication bidirectionally from a fixed point—the *origin.*

Ribosomes

Ribosomes are distributed throughout the cytoplasm and are the sites of protein synthesis. Composed of RNA and proteins: organized in two sub-units 30s and 50s.

Cytoplasmic inclusions

Sources of stored energy, e.g. polymetaphosphate (volutin), poly-β-hydroxybutyrate (lipid), polysaccharide (starch or glycogen).

Spores

Spores produced by bacteria in the genera *Bacillus* and *Clostridium* enable them to survive adverse environmental conditions: developed from and at the expense of the vegetative cell. Spores are dense, contain a high concentration of calcium dipicolinate and are resistant to heat, desiccation and disinfectants; they often remain associated with the cell wall of the bacillus from which they develop and are described as 'terminal', 'subterminal', etc. When growth conditions become favourable they germinate to produce vegetative cells.

TAXONOMY

Taxonomy is the classification or division of organisms into ordered groups.

Nomenclature is the labelling of the groups and of individual members within groups.

Organisms fall into three *kingdoms:*
1. Animals
2. Plants
3. Protista—contains all unicellular organisms including bacteria.

Higher organisms are classified phylogenetically (i.e. on the basis of evolution) but this classification is impossible in the case of bacteria as they lack sufficient morphological features.

Bacterial classification is therefore artificial: different characteristics have been chosen arbitrarily so that the various members can be distinguished. These characteristics include:

Morphology
Staining
Cultural characteristics
Biochemical reactions
Antigenic structure
Base composition (i.e. GC ratio) of bacterial DNA

Gram-positive bacteria

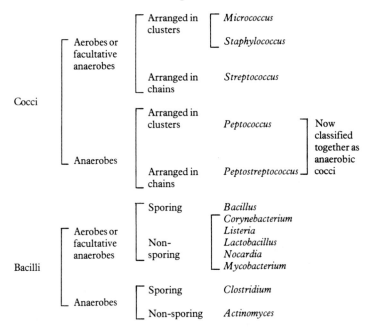

Fig. 2.6 Simplified classification of Gram-positive bacteria.

Figures 2.6 and 2.7 list most of the medically important bacterial genera classified on the basis of Gram's stain, morphology and aerobic or anaerobic growth: note that this classification has been simplified — for example, microaerophilic organisms are classified as anaerobes.

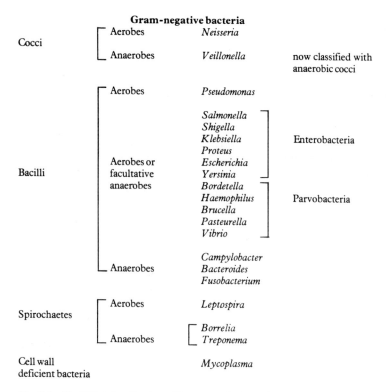

Fig. 2.7   Simplified classification of Gram-negative bacteria.

3

Growth and nutrition of bacteria

Bacteria, like all cells, require nutrients for the maintenance of their metabolism and for cell division. Fast-growing bacteria divide approximately every 30 mins.

Chemically, bacteria consist of:
Protein
Polysaccharide
Lipid
Nucleic acid
Peptidoglycan

Bacterial growth requires:
1. materials for the synthesis of structural components and for cell metabolism
2. energy

Bacteria differ widely in their nutritional requirements. Some bacteria can synthesise all they require from the simplest elements. Others—including most pathogenic bacteria—are unable to do this: they need a ready-made supply of some of the organic compounds required for growth; other necessary compounds can be synthesised from breakdown products of complex macromolecules (e.g. proteins, nucleic acids) which are taken into the cell and degraded by bacterial enzymes. This is illustrated diagrammatically in Figure 3.1.

Elements. Bacterial structural components and the macromolecules for cell metabolism are synthesised from the elements shown in Table 3.1; all are therefore necessary for bacterial growth—whether available simply as elements or as part of complex molecules.

The four most important elements are:
1. Hydrogen
2. Oxygen
3. Carbon
4. Nitrogen

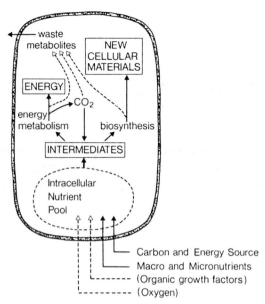

Fig. 3.1　Bacterial nutrition and metabolism

Hydrogen and oxygen

These are obtained from water—essential for the growth and maintenance of any cell: water must therefore be available in any situation of potential bacterial growth.

Carbon and nitrogen are the principal elements for which an external source must be found.

Carbon

Depending on their requirements, bacteria can be classified as:

1. *Autotrophs:* free-living, non-parasitic bacteria, most of which can use *carbon dioxide* as their carbon source. The energy needed for their metabolism can be obtained from:
   a. Sunlight—photoautotrophs
   b. Inorganic compounds by oxidation—chemoautotrophs
2. *Heterotrophs:* generally parasitic bacteria; require more complex, *organic compounds* than carbon dioxide as their source of carbon and energy, e.g. sugars.

Human pathogenic bacteria are heterotrophs: they are parasitic and

Table 3.1   Essential elements required for bacterial growth

Group	Element	Required for:
I Elements required for the synthesis of structural components	Carbon Hydrogen Oxygen Nitrogen Phosphorus Sulphur	Synthesis of: Carbohydrate and lipid, protein, nucleic acid
II Elements required for other cellular functions	Potassium	Major cation; activates various enzymes
	Calcium	Enzyme cofactor (e.g. proteinases): key role in spore formation
	Magnesium	Multi-enzyme cofactor, stabilises ribosomes, membranes, nucleic acid; enzyme substrate binding
	Iron	Electron carrier in oxidation-reduction reactions; many other functions
III Trace elements	Copper Cobalt Manganese Molybdenum Zinc	Activators and stabilisers of a wide variety of enzymes

have evolved to adapt to an environment—i.e. the human body—in which many of their nutrients are ready-made in complex form and freely available. Such bacteria have lost the biosynthetic mechanisms necessary for a free-living existence in a harsher environment, e.g. soil or water.

The principal source of carbon is carbohydrate (usually sugars).

Sugars are degraded either by

a) *oxidation* or

b) *fermentation* (i.e. without oxygen) This provides energy for the generation of adenosine triphosphate (ATP) the universal energy storage compound.

Nitrogen

The main source of nitrogen is ammonia, usually in the form of an ammonium salt: this is available either in the environment or produced by the bacterium as a result of deamination of amino acids released from proteins.

Organic growth factors

These are organic compounds that cannot by synthesised by many bacteria; an exogenous supply is therefore required, although often only in small amounts.
Some examples are listed below:

1. *Amino acids:* required by many bacterial species which cannot synthesise them from simpler elements. Bacteria possess enzymes that degrade proteins to amino acids; these then form an intracellular pool from which the appropriate amino acid is withdrawn to become incorporated into bacterial proteins.
2. *Purines and pyrimidines:* the precursors of nucleic acids and coenzymes. In bacteria which require them, they are converted into nucleosides and nucleotides before incorporation into DNA and RNA.
3. *Vitamins.* Many pathogenic bacteria lack the ability to synthesise vitamins—most of which are required for the formation of coenzymes.

Prototrophs and auxotrophs

Prototrophs are *wild-type* bacteria with 'normal' growth requirements. *Auxotrophs* are mutants which require an additional growth factor not needed by the parental or wild-type strain: the growth factor may be different types of chemical compound, e.g. an amino acid, pyrimidine base, vitamin.

NUTRIENT UPTAKE

Most nutrients are small molecules which diffuse freely across the bacterial cytoplasmic membrane to enter the cell: some are at a higher concentration within the bacterial cell than in the external environment so that their uptake is an energy-dependent process.

Sugars are nutrients of relatively large size and therefore diffuse slowly into the bacterial cell.

Enzymes which facilitate the rapid uptake of larger nutrient molecules are present in many bacteria: usually associated with the cell membrane and energy-dependent, they may be

1. *Inducible*—i.e. produced only in the presence of the substrate *or*
2. *Constitutive*—i.e. produced constantly and independently of the substrate.

ENVIRONMENTAL CONDITIONS GOVERNING GROWTH

Water

Moisture is an absolute requirement for the growth of all bacteria: at least 80 per cent of the bacterial cell consists of water: the availability of water, for example, largely determines the size of the population of bacteria that can be supported by the skin.

Oxygen

Bacteria differ in their need for molecular oxygen for growth or—in the case of anaerobic bacteria—in their need for its exclusion: this is illustrated in Table 3.2.

Table 3.2 Effect of oxygen on bacterial growth

Bacteria	Growth	
	In free oxygen	In absence of oxygen
Aerobes		
strict aerobes	+	−
facultative anaerobes	+	+
Anaerobes		
strict anaerobes	−	+
microaerophiles	−	+ (can grow with trace of both oxygen and carbon dioxide)

Carbon dioxide

Required by all bacteria and usually available as a product of metabolism. Slow-growing or fastidious organisms may not generate enough carbon dioxide so that this must be supplied exogenously: this requirement becomes increased by environmental change e.g. caused by the transfer of bacteria from growth *in vivo* to culture *in vitro*.

Many pathogenic bacteria therefore require addition of 5 to 10 per cent carbon dioxide to the incubator atmosphere for *primary isolation* in vitro from clinical material.

Temperature

Bacteria also differ with regard to the optimal temperature range for their growth:

Psychrophile   below 20 °C
Mesophile   between 25 °C and 40 °C
Thermophile   between 55 °C and 80 °C

Most medically important species are mesophiles and grow best at temperatures around 37 °C (i.e. body temperature).

Hydrogen ion concentration

Not surprisingly, the optimal pH for bacteria that have evolved in association with man is similar to physiological pH—i.e. 7.2 to 7.4. A few species have evolved to become adapted to ecological niches where the pH is either higher or lower than normal.

BACTERIAL GROWTH AND DIVISION

Bacterial growth is the result of a balanced increase in the mass of cellular constituents and structures; the biosynthetic processes on which this increase depends are fuelled by energy usually from ATP.

Cell division is initiated when the increase in cellular constituents and structures reaches a critical mass.

Bacteria divide by *binary fission.*

Bacterial chromosome: a circular double-stranded DNA molecule. It replicates semiconservatively and bidirectionally: the replicated genome *segregates* between the daughter cells; cross walls then form as the parent cell divides into the two daughter cells.

Bacterial growth cycle

The growth cycle of a bacterial population (on transfer into fresh medium) is shown in Figure 3.2.

Four main phases can be recognised:
1. *Lag phase:* bacteria do not divide immediately but initially undergo a period of adaptation with active macromolecular synthesis.
2. *Exponential (log) phase:* cell division then proceeds at a logarithmic rate determined by the medium and conditions of the culture.

During the exponential phase the population can double approximately every 30 min (with fast-growing bacteria): this is known as the *doubling time,* or *mean generation time,* and it can be calculated from the slope of the plot of the growth curve.

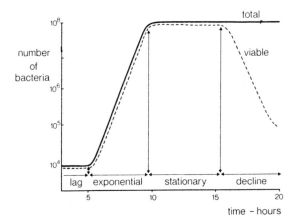

Fig. 3.2   The bacterial growth cycle.

3. *Stationary phase:* is reached when one or more essential nutrients become depleted; cell division ceases and there is no further growth: represented by the plateau in Fig. 3.2 when the bacteria have achieved their *maxinal cell density* or yield.
4. *Decline phase:* after a period in the stationary phase, the bacteria start to die although the number of cells (viable and non-viable) remains constant.

Growth in vivo

Bacterial growth in the human body is very different from that in artificial culture in the laboratory. In general, the growth rate in vitro is much faster.

Various factors influence bacterial growth in vivo:
1. Mean bacterial generation time
2. Nutritional status of patient
3. Redox potential (E_h)
4. Hydrogen ion concentration (pH)
5. Presence of metals e.g. iron, calcium
6. Localisation of nutrients
7. Cellular defences (reticulo-endothelial system)
8. Humoral defences (immunoglobulin, complement)
9. Host enzymes (proteases, hyalurenidase)

Some bacteria need special growth factors e.g. the growth of *Brucella abortus* is enhanced by erythritol (a substance found in abundance in placentae where *Br. abortus* multiplies preferentially).

4

Bacterial genetics

Genetics is the study of inheritance. Except for RNA viruses, all inherited characteristics are encoded in DNA.

Bacteria have two types of DNA that contain their genes:

1. Chromosomal
2. Extrachromosomal—i.e. plasmid

THE BACTERIAL CHROMOSOME

Bacteria are prokaryotes in that their chromosome is not contained with a nuclear membrane (unlike eukaryotes). The chromosome is

1. Circular, double-stranded DNA
2. Attached to the bacterial cell membrane (Fig. 4.1).

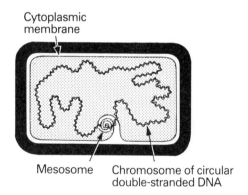

Fig. 4.1   Diagram to show bacterial genome.

Genetic code is contained in the sequence of purine and pyridimidine bases of the nucleotides which make up the DNA strand. Three bases comprise one codon and each triplet codon codes for one amino acid.

In this way the gene determines the sequence of amino acids which form the protein which is the gene product.

DNA replication is semi-conservative, i.e. each strand of DNA is conserved intact during replication and becomes one of the two strands of the new daughter molecules (Fig. 4.2).

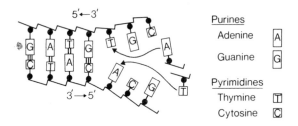

Fig. 4.2   Diagram to show bidirectional replication of bacterial DNA.

DNA-dependent DNA polymerase is the main replicase enzyme involved in DNA replication although many other enzymes take part in the process.

Repair mechanisms exist in bacteria: these excise incorrect nucleotide sequences with nucleases, replace them with the correct nucleotides and religate the sequence.

Restriction enzymes are endonucleases and are a type of defence mechanism found in many bacteria against incoming foreign nucleic acids. They cleave double-stranded DNA at specific sequences (usually of about six nucleotides). The pattern of fragments produced can be demonstrated by gel electrophoresis and is reproducible and constant.

PLASMIDS

Plasmids are extrachromosomal DNA molecules. Smaller than the chromosome, they also consist of circular, double-stranded DNA but most often within the size range 10–200 megadaltons.

Replication is autonomous: plasmids multiply independently of the host cell (i.e. they are replicons) but also divide with the cell so that they are inherited by daughter-cells.

Multiple copies of the same plasmid may be present in each bacterial cell.

Different plasmids are also often present in the same cell. Usually one or two plasmids (but occasionally as many as eight) different plasmids may co-exist within the same bacterium.

Transmissibility: some plasmids (but not all) can transfer to other bacteria of the same and also of different species. Transfer takes place normally by conjugation and the ability to transfer is mediated by the *tra* or transfer promotion genes. Some plasmids cannot transfer themselves but can nevertheless be transferred along with other *tra*⁺ plasmids.

Maintenance of the plasmid in the cell also requires the expression of other plasmid genes.

Plasmid types

There are many types of plasmid. Below are some examples:

1. *R-factors:* plasmids which contain genes that code for antibiotic resistance (Fig. 4.3).

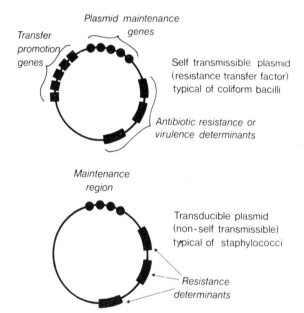

Fig. 4.3 The structure of two types of plasmid that code for antibiotic resistance in bacteria.

2. *Col factors:* found in many species of enterobacteria, produce extracellular toxins (colicines) that inhibit strains of the same and different species of bacteria.
3. *F or fertility factor:* This much-studied plasmid is a useful tool for mapping genes on the bacterial chromosome because it can itself

recombine with the bacterial chromosome. It then promotes transfer of the chromosome at a high frequency of recombination (Hfr) into the chromosome of a second (recipient) bacterial cell during mating. The chromosome transfers as a linear molecule, the F factor going over last.

By stopping transfer at different times after mating, the genes of the first bacterial chromosome can be mapped. This is done by determining which bacterial gene function or products can be detected after different transfer times in the recipient bacterium.

4. *Genetic engineering:* multi-copy plasmids with a high rate of genetic expression are widely used as vectors in genetic engineering. Genes that code for certain proteins (such as interferon) are ligated into the DNA of the plasmid vector and introduced into cells such as *Esch.coli* or yeasts where they are expressed very efficiently.

GENETIC VARIATION IN BACTERIA

This can take place by:
1. Mutation
2. Gene transfer

Mutation

Mutation is due to a chemical alteration in DNA. Mutants are variants in which one or more bases in their DNA are changed: this change is heritable and irreversible (unless there is back-mutation to the original sequence). The gene defect may result in:
1. Alteration in the process of transcription.
2. Alteration in the amino-acid sequence of the protein which is the gene product.

Mutation can involve any of the numerous genes in bacterial DNA: many mutations are never detected because detection depends on the mutation affecting a recognisable function (e.g. causing antibiotic resistance).

Molecular basis of mutation

Mutation involves change in the sequence of bases in DNA. Change of a single base alters the genetic code so that the triplet involved codes for a different amino acid; the new amino acid then becomes substituted for the correct amino acid in the protein product of the gene affected.

Mutation can be of three types:

1. *Base substitution:* change of a single base to one of the three other bases with consequent alteration in the triplet of the code. This may be:
   a. *Transition* in which purine/pyrimidine orientation is preserved, e.g. GC changes to AT
   b. *Transversion* with altered purine/pyrimidine orientation, e.g. GC changes to CG.
2. *Deletion:* loss of a base to affect the reading of subsequent triplets — frame-shift mutation (see Fig. 4.4). Deletion sometimes involves several bases rather than a single base.
3. *Insertion* of an additional base also alters the reading frame of the DNA (Fig. 4.4).

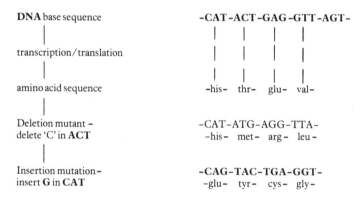

Fig. 4.4 Mutation. The effect of the deletion and insertion of a single base on the amino acid sequence of the gene product.

Gene transfer

There are three types of gene transfer that alter the DNA gene content of bacteria:
1. Transformation
2. Transduction
3. Conjugation

Transformation

Transformation is when fragments of exogenous bacterial DNA are taken up and 'absorbed' into recipient cells.

Recombination with the bacterial chromosome takes place (Fig. 4.5) to *transform* the cell which then expresses the new genes. Transformation is detected by an alteration in the behaviour and characteristics of the recipient bacteria. As in other bacterial systems, recombination depends on extensive DNA homology and on the function of a gene known as the *recA* gene.

Transformation is illustrated diagramatically in Figure 4.5.

Bacterial chromosome

Exogenous fragment of bacterial DNA

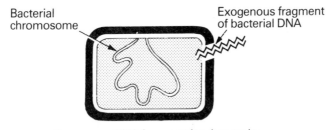

Exogenous DNA fragment is taken up by 'competent' recipient bacterium

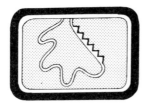

One strand of incoming DNA recombines with region of homology in recipient bacterial chromosome

Fig. 4.5 Transformation. Gene transfer by the uptake and subsequent recombination of a fragment of exogenous bacterial DNA.

Competence: The recipient bacteria must be competent—usually a transitory state in microbial cultures.

Frequency: The frequency of transformation is low.

Note: Transformation can be carried out with either chromosomal or plasmid DNA. It is a laboratory manipulation and probably does not take place in nature.

Transduction

Fragments of chromosomal DNA can be transferred or transduced into a second bacterium by phage. During phage replication, a piece of bacterial DNA becomes, by accident, enclosed within a phage particle instead of the normal phage DNA (Fig. 4.6). When this particle infects

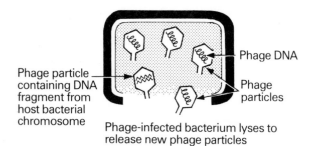

Phage particle containing DNA fragment from host bacterial chromosome

Phage DNA

Phage particles

Phage-infected bacterium lyses to release new phage particles

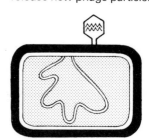

Phage carrying bacterial DNA fragment enters recipient bacterium

Bacterial DNA from phage recombines into chromosome of recipient bacterium

Fig. 4.6   Transduction. Gene transfer from one bacterium to another via phage.

a second bacterial host cell, the DNA from the first bacterium is released and becomes recombined into the chromosome of the second bacterium. The bacterial genes transferred in this way are therefore expressed.

Plasmid DNA can also be transferred to the second bacterium by transduction. The donated plasmid can then function (and replicate) independently, i.e. without recombining into the chromosome of the bacterial cell.

Beta lactamase production: in *Staphylococcus aureus* is plasmid mediated and the responsible plasmids transfer between staphylococcal strains by transduction.

Transduction otherwise seems to be a relatively uncommon event in nature.

Phage conversion. Phage DNA (as distinct from bacterial DNA) itself becomes integrated into the bacterial chromosomel. The phage genes cause changes in the phenotype of the host bacterium for example, toxin production in *Corynebacterium diphtheriae* and the production of certain O antigens in salmonellae. Integration of phage DNA into the bacterial genome acts as a switch to cause expression of otherwise unexpressed bacterial genes. However, when the phage is lost from the bacterium so, of course, are the new characteristics. In other words, the continuing presence of the phage DNA is necessary for the maintenace of the altered phenotype.

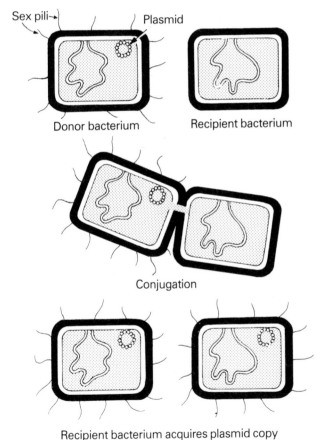

Fig. 4.7 Plasmid gene transfer by conjugation.

Conjugation

Conjugation is the major way in which bacteria—particularly entero-bacteria—acquire additional genes. In conjugation, plasmid DNA is transferred from donor to recipient bacterium by direct contact probably via a hollow-cored tube formed by a sex pilus (Fig. 4.7). The formation of the pilus is coded by plasmid *tra* genes.

TRANSPOSONS

Sometimes called 'jumping genes', transposons are segments of DNA that can transpose or move extremely readily, from plasmid to plasmid or from plasmid to chromosome (and vice versa). In this way, plasmid genes can become part of the chromosomal complement of genes. When transposons transfer to a new site, it is usually a copy of the transposon that moves, the original transposon remaining in situ. Transposons are insertion sequences and consist of a unique sequence of DNA bounded by terminal inverted or sometimes direct repeated DNA sequences (Fig. 4.8); the repeat sequences are responsible for the ability to translocate and insert into DNA.

For their insertion, transposons do not require extensive homology between the terminal repeated sequences of the transposon (which are responsible for integration) and the site of insertion in the recipient DNA, although certain sites are preferred. Unlike classical recombination in bacteria, transposition is independent of the function of the *recA* gene.

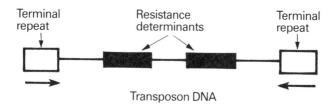

Fig. 4.8   Transposon. Diagram to show structure of transposon DNA.

Coding. Transposons code for toxin production, resistance to several antibiotics, e.g. ampicillin, trimethoprim as well as other functions. Some examples of transposons which mediate antibiotic resistance are listed in Table 4.1

Transposition is a mechanism which allows a high rate of change in the genes contained in plasmids and bacterial chromosomes.

Table 4.1   Some transposons and the antibiotic resistance that they encode

Transposon (Tn)	Antibiotic resistance
Tn1	Ampicillin
Tn4	Ampicillin, streptomycin, sulphonamide
Tn5	Kanamycin
Tn7	Trimethoprim, low level streptomycin
Tn9	Chloramphenicol
Tn10	Tetracycline

Note, however, transposons do not exist in the free state but only in the integrated form within plasmid or chromosome. They are not themselves, therefore, capable of mediating gene transfer between bacterial cells.

5

Laboratory Methods

Bacterial disease is diagnosed principally by culture of the organisms responsible. Special techniques are necessary for this and for the microscopical demonstration of organisms together with their identification both biochemically and serologically.

Below are the five main categories of methods used in the laboratory:

1. Microscopy
2. Culture
3. Bacterial identification
4. Tests of antimicrobial drugs
5. Serology

MICROSCOPY

Preparations

1. *Stained film:* heat-fixed smears on slides of specimens or bacterial cultures stained by flooding with appropriate dyes.
   Observe: staining characteristics, shape, arrangement of bacteria, presence of cells, e.g. polymorphs (pus cells).
2. *Unstained (wet) films:* drops of liquid specimens or fluid cultures placed on a slide and covered with a coverslip.
   Observe: for bacteria and cells, bacterial motility.

Note: the refractive index of bacteria is similar to that of the fluids in which they are suspended and can only be seen by cutting down the illumination of the standard light microscope.

Microscopes

1. *Standard light microscopy:* the main method used in diagnostic bacteriology.

Magnification:

a. *Stained films:* are examined with the oil immersion objective (×100): using ×10 eyepieces the final magnification is ×1000.

b. *Wet films:* are examined with high power dry objective (×40): final magnification with ×10 eyepieces is ×400.

Definition: the ability to render the outline of the object clear and distinct depends on the quality of the lenses: good (but expensive) optical systems produce less aberration.

Resolution: the ability to distinguish two adjacent points as separate entities is determined by the wave length of light: the best resolution obtainable is of the order of 0.23 μm.

2. *Dark-ground microscopy:* a special condenser illuminates the specimen obliquely so that light does not enter the objective. Light is scattered by bacteria which appear brightly illuminated against a dark background.

3. *Phase contrast microscopy:* a special condenser and objective are used: direct light from the source and light scattered by structures in the field are transmitted so that when direct and diffracted beams unite they are not in phase: details of the object appear as differences in intensity and the contrast produced shows details of the fine structure of unstained, living microorganisms.

4. *Fluorescence microscopy:* this uses ultraviolet light. Bacteria or cells stained with auramine or other suitable fluorescent dyes alter the wave length and become visible as bright objects against a dark ground.

Immunofluorescence: this combines serology with fluoresence microscopy by using antibody labelled with fluorescent dyes (e.g. fluorescein isothiocyanate, lissamine, rhodamine) to detect specific antigens and so identify bacteria.

5. *Electron microscopy:* a beam of electrons allows resolution of extremely small objects e.g. 0.001 μm: the electrons are focused by electro-magnetic fields and the image is visualised—and can be photographed—on a fluorescent screen: used for study of the ultrastructure of bacteria.

Staining

1. *Gram's stain:* the most widely used stain in medical bacteriology: it not only reveals the shape and size of bacteria but enables them to be immediately classified into two categories—Gram-positive and Gram-negative.

Method: crystal violet, iodine solution, decolourise with acetone or alcohol, counterstain (e.g. dilute carbol fuchsin).

Observe:

a. *Gram-positive* bacteria resist decolourisation and stain blue-black.
b. *Gram-negative* bacteria are decolourised and so stain pink with the counterstain.
c. *Cells* are Gram-negative: Gram stain shows sufficient detail for polymorphs to be distinguished from other cells but does not reveal much cytological detail.

2. Staining for acid and alcohol-fast bacilli (usually tubercle bacilli).
a. *Ziehl-Neelsen method:* concentrated carbol fuchsin (heated), decolourise with acid *and* alcohol, counterstain with methylene blue or malachite green.

Observe: for red bacilli against a blue ground.

b. *Auramine method:* auramine-phenol, decolourise with acid *and* alcohol, apply potassium permanganate.

Observe: for fluorescent yellow bacilli in a dark field under ultra-violet light.

The auramine method is now preferred by many laboratories.

Medical bacteriologists use various special stains in addition to those listed above, e.g. for demonstration of volutin granules (Albert's or Neisser's stain), capsules, spores, spirochaetes, flagella etc.

CULTURE

Bacteria grow well in vitro on artificial media. However, although they differ in their growth requirements so that many different kinds of media must be used, the majority of pathogenic bacteria grow on blood agar—the mainstay of diagnostic bacteriology.

Growth requirements

Most pathogenic bacteria in microbiology are heterotrophs: their nutritional and metabolic requirements are sophisticated and they need organic materials (e.g. carbohydrates, amino acids) and sometimes specialised growth factors for culture in the laboratory.

Media

Constituents of culture media

1. Water
2. Sodium chloride, other electrolytes
3. 'Peptone': a protein digest prepared from animal or vegetable protein by enzymic action; contains peptones, proteases, amino acids, etc.

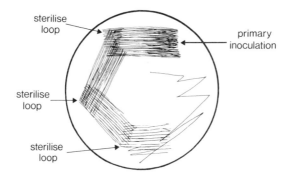

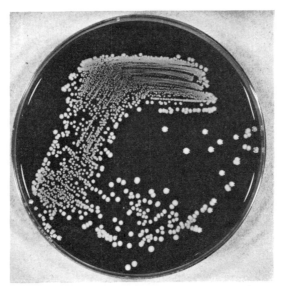

Fig. 5.1 Plating. Diagram illustrating the method of inoculating a plate of solid medium with bacteria to achieve separated colonies and a blood agar plate inoculated in this way after overnight incubation.

4. Meat extract, yeast extract: used to enrich media; contain protein degradation products, carbohydrates, inorganic salts, growth factors.
5. Blood: usually defibrinated horse blood: sometimes serum from other animals.
6. Agar: a carbohydrate derived from sea weeds; its unique property is that it melts at 90 °C but does not solidify until cooled to 40 °C. Heat-sensitive ingredients can therefore be added just before the medium sets.

Nowadays media are prepared from dehydrated ingredients supplied by commercial firms: these are simply reconstituted and sterilised in the laboratory before use.

Solid media

Dispensed in plastic or glass petri dishes, solid media are solidified fluid media.

Table 5.1 Solid media

Medium	Main constituents	Use
Nutrient agar	Nutrient broth agar	General culture
Blood agar	Nutrient agar 5 to 10% horse blood	General culture; the most widely used medium in medical bacteriology
Chocolate agar	Heated blood agar	Isolation of *H. influenzae, N. gonorrhoeae*
MacConkey agar	Peptone water agar, bile salt, lactose, neutral red	Culture of enterobacteria: lactose-fermenting colonies are coloured pink
CLED agar (cystine-lactose electrolyte deficient medium)	Peptone, L-cystine, lactose, bromthymol blue	Culture of enterobacteria: lactose-fermenting colonies are coloured yellow
Desoxycholate citrate agar	Nutrient agar, sodium desoxycholate, sodium citrate, lactose, neutral red	Selective medium for salmonellae, shigellae
Lowenstein-Jensen*	A mineral salt solution, glycerol, malachite green, whole egg	Culture of *Myco. tuberculosis*
Antibiotic sensitivity ('Isosensitest')	Peptone in a semi-synthetic medium designed to avoid antibiotic inhibitors	Antibiotic sensitivity tests

*Note: This is rendered solid by heating which 'sets' the egg: it does not contain agar.

Agar: the setting agent, added in a concentration of 1.5 per cent to give the medium the consistency of a firm jelly.

Method of inoculation: specimens or cultures of bacteria are plated or stroked out on the medium with a wire loop in such a way as to ensure a reducing inoculum: this means that after incubation, separated colonies will develop where individual bacteria have been deposited. (Fig. 5.1).

The most widely used are shown in Table 5.1.

Advantages and uses

1. The great value of a solid medium is that it allows separate colony formation.
2. *Colonial morphology* enables most bacterial species to be presumptively identified.
3. *Quantitation:* the numbers and relative proportion of different bacterial species originally present in the specimen can be estimated.
4. *Pure cultures* can be obtained by picking isolated bacterial colonies onto fresh solid medium—necessary for full identification.

Selective media are solid media containing ingredients which inhibit

Table 5.2 Liquid media

Medium	Main constituents	Use
Peptone water	Peptone, sodium chloride, water	General culture; basal medium for sugar fermentation tests
Nutrient broth	Peptone water, meat extract	General culture
Glucose broth	Nutrient broth, glucose	Culture of delicate organisms
Robertson's meat medium	Nutrient broth, minced meat	Culture of anaerobic and aerobic bacteria
Tetrathionate broth	Nutrient broth, sodium thiosulphate, iodine	Enrichment culture for salmonellae
Selenite F broth	Peptone water, sodium selenite	Enrichment culture for salmonellae and shigellae

unwanted contaminants (e.g. from the normal flora) but allow certain pathogens to grow.

Liquid media

Dispensed in tubes with cotton-wool stoppers or screw-capped bottles.

Growth is recognised: by turbidity in the fluid. The most common are listed in Table 5.2.

Uses

1. Some bacteria, particularly if present in small numbers, will grow only in fluid media: occasionally a specimen contains inhibitory substances e.g. antibiotics, which are diluted by inoculation into a fluid medium thus allowing bacterial growth.
2. Fluid media with special constituents (e.g. sugars) are widely used to test the biochemical activities of bacteria for identification.
3. Enrichment media are fluids which encourage the preferential growth of a particular bacterium: they contain inhibitors for contaminants which might otherwise overgrow the pathogen.

Disadvantages

1. Identification of bacteria by colonial morphology is not possible.
2. No estimate can be made of the numbers or the relative proportion of different species of bacteria originally present in the specimen.

Media for blood cultures. Two bottles with rubber seals and a perforated metal cap: inoculated by injection of aseptically-collected blood through the hole in the cap with a syringe. They contain:

1. For aerobic culture: nutrient broth sometimes with an agar slope set onto the narrow side of the bottle so that colonies can be observed.
2. For anaerobic culture: nutrient broth with sodium thioglycollate (a reducing agent) alternatively Robertson's meat medium:

Transport medium. For the preservation of delicate pathogens during transit to the laboratory. For example, *Stuart's Transport Medium,* a semi-solid, non-nutrient agar with thioglycollic acid (as reducing agent) and electrolytes. Originally devised for *Neisseria gonorrhoeae* but now used as a general transport medium.

Incubation

1. *Atmosphere*

   a. Most human pathogens grow in air but, the addition of 10 per cent carbon dioxide is essential for the isolation of some species and enhances the growth of many others.

   b. *Anaerobic bacteria* require incubation without oxygen: plates of media are inoculated and are placed in a sealed jar from which oxygen is removed. Hydrogen and carbon dioxide are liberated inside the jar by means of a commercially available gas generating system: water forms by combination of the oxygen with the hydrogen in the presence of a catalyst (e.g. alumina pellets coated with palladium).

2. *Temperature*

   The optimal temperature for the growth of most pathogens is body heat—37 °C: a few bacteria require a higher and some a lower temperature.

Table 5.3   Biochemical tests for bacterial identification

Test	Substrate	Observe

Carbohydrate metabolism
Breakdown of sugars
  a. Oxidative *(aerobic)*
  b. Fermentative
    (anaerobic)
	Various carbohydrates	Acid, sometimes with gas

Voges-Proskauer	Glucose	Acetyl methyl carbinol
Citrate utilisation	Sodium citrate	Growth (citrate is the sole carbon source)
Protein metabolism Gelatin liquefaction	Gelatin	Liquefaction
Amino acid decarboxylases	Lysine, ornithine, arginine	Carbon dioxide
Production of hydrogen sulphide	Sulphur-containing amino acids	Hydrogen sulphide
Phenylalanine deaminase	Phenylalanine	Phenylpyruvic acid
Indole	Tryptophan	Indole
Other tests Urease	Urea	Ammonia
Catalase	Hydrogen peroxide	Oxygen
Oxidase	Redox dye	Oxidisation to purple colour
Nitrate reduction	Potassium nitrate	Nitrate

Note. Most of these tests are devised so that the products result in the development of a visible change e.g. indicators for pH change, gas production etc.

BACTERIAL IDENTIFICATION

Bacteria isolated by culture from a specimen must be identified. This proceeds in the following stages:

1. Colonial and microscopic morphology

The appearance of bacterial colonies is usually characteristic and colonies of most pathogenic bacteria can be recognised, at least presumptively, on the primary cultures. A stained film of the colony may be required for identification and in any event most bacterial species require confirmatory tests.

2. The conditions required for growth

These also help to identify bacteria, e.g. whether aerobic or anaerobic growth is required, ability to grow on simple or only on enriched or selective media etc.

3. Biochemical tests

The main tests are listed in Table 5.3; these are principally used for enterobacteria.

API system: nowadays most laboratories use commercially prepared kits with which a wide range of tests can be carried out rapidly and accurately: the best are the API systems (Fig. 5.2).

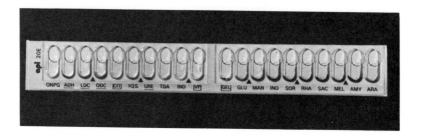

Fig. 5.2 API strip. A commercial kit of 20 biochemical tests used to identify enterobacteria. Also available for various other bacteria.

Method: after inoculation and incubation the results are scored to give a numerical profile. This is compared statistically to a profile compiled from type cultures and the degree of correspondence enables identification to be made with a known degree of certainty.

4. Recognition of enzymes

Although enzyme production is the basis of most of the reactions included in the biochemical tests listed in Table 5.3, some bacteria can be identified primarily by production of a characteristic enzyme. For example:

a. *Coagulase:* clots plasma (specific for *Staphylococcus aureus*)

b. *Lecithinase:* produces opacity in egg or serum medium (if inhibited by *Clostridium perfringens* antitoxin, identifies this organism—the Nagler Reaction).

5. Antigenic structure

Serology is the definitive method of identification. In bacteriology it depends mainly on recognition of antigens in flagella, cell wall, capsule or liberated from the bacteria as toxins. Particularly useful for the large numbers of biochemically-similar enterobacteria (e.g. Salmonellae).

Methods: various immunological techniques are used, most often agglutination, but also precipitation, sometimes neutralisation of toxin.

6. Typing of bacterial strains

To trace epidemic spread of an organism, it is often necessary to identify individual strains or types within a bacterial species. This can be done in the following ways:

a. *Antigenic typing:* to distinguish different strains of bacteria by their antigenic structure.

b. *Bacteriophage typing:* strains of an organism differ in their susceptibility to a series of bacterial viruses or bacteriophages: patterns of lysis with more than one phage are usually detected.

c. *Bacteriocine typing:* bacteriocines (see Ch. 4) are proteins released by bacteria which inhibit the growth of other members of the same species: strains can be differentiated on the basis of the bacteriocins they produce by observing the patterns of inhibition of a series of test strains.

TESTS OF ANTIMICROBIAL DRUGS

Sensitivity tests

One of the most important functions of a diagnostic bacteriology laboratory and a large part of the day-to-day workload.

There are two principal methods:

1. Disc diffusion

By far the most widely used method.

Method: papers discs impregnated with antibiotic solutions (at a concentration related to blood or urine levels) are placed on the surface of a plate inoculated all over with either the specimen or the bacterial culture under test. Fig. 5.3.

Fig. 5.3 Antibiotic sensitivity test showing zones of inhibition of the growth of the test organism round discs containing antibiotics to which it is sensitive. Resistance is shown by the growth of the orgamism right up to the disc.

Stokes' method: the outside of the plate can be inoculated with a standard organism (e.g. the Oxford strain of *Staphylococcus aureus*); the zones round the discs can then be compared to those produced against the test organism in the middle of the plate. Fig. 5.4.

Fig. 5.4 Antibiotic sensitivity test—Stokes' method. The fully sensitive test organism is on the outside. Reduced zones around two of the discs indicate resistance; in some tests there are no zones at all.

Observe: for zones of inhibition round the discs after incubation.

Primary sensitivity testing: the specimen is inoculated directly onto a sensitivity plate. Often successful with infected urines, it may fail to give a readable result with other specimens.

Secondary sensitivity testing: the inoculum is the bacterium isolated from the specimen: often preferred and is essential when the growth is mixed. Unfortunately, it involves a day's delay in reporting.

2. Tube dilution

Laborious: only done in special circumstances e.g. in tests on bacteria from cases of infective endocarditis where treatment is difficult.

Method: tube with doubling dilutions of the antibiotic in broth are inoculated with the bacterium under test.

Observe: for tubes in which bacterial growth is inhibited after incubation; the lowest concentration in which there is no growth represents the MIC or minimum (bacteriostatic) inhibitory concentration.

MBC or minimum bactericidal concentration is estimated by subculture from the tubes of a MIC test onto solid media; growth on subculture indicates the presence of surviving bacteria and that the concentration has not been bactericidal. Estimated as the lowest concentration of antibiotic in which the bacteria have been killed (i.e. no growth on subculture from the appropriate tube).

Tests of combined antibacterial action

Carried out by the tube dilution method but using combinations of two antibiotics in concentrations which correspond to blood levels; usually only done when indicated clinically e.g. in infective endocarditis due to a resistant organism. The dilutions may be tested in a 'chessboard' format to demonstrate synergy or, possibly, antagonism.

Estimation of level of antimicrobial agents in blood

1. To check that the concentration of drug in the blood is below that associated with toxicity
2. To confirm that the blood contains an adequate therapeutic level of the drug.

Method: specimens of serum are taken before and after a dose to estimate 'trough' and 'peak' levels respectively. They can be tested in one of two ways:

a. *Immunochemical assays:* fast, accurate and now widely used. Tests are run on computerised assay equipment with commercially prepared kits. Still only available for a few antibiotics—principally aminoglycosides.

b. *Bioassay:* measuring zones of inhibition produced by serum in wells in an agar plate inoculated with a suitable bacterial culture: these are compared to zones produced by standard concentrations of the drug and the level in the serum calculated.

SEROLOGY

Some diseases—fewer now than formerly—are diagnosed by demonstration of antibody to the causal organism in the patient's blood.

Titre: is the term for the highest dilution of serum at which antibody activity is demonstrable, usually expressed as the reciprocal of the serum dilution e.g. 64 if antibody was detected at a final serum dilution of 1 in 64.

Methods

The main immunological techniques for antibody detection are:

1. Agglutination

Antibody in patient's serum is detected by its ability to cause visible aggregation of suspensions of bacteria.

Indirect (Coombs) agglutination: sometimes the antibody in a patient's serum is incomplete and although it combines with bacteria does not cause them to agglutinate: later addition of rabbit anti-human globulin causes the antibody-coated bacteria to agglutinate.

2. Precipitation

The antigen is in soluble form and antibody is detected usually by formation of a visible line of precipitate as a result of diffusion in agar gels.

3. Complement fixation

Combination of antigen with antibody 'fixes' or uses up complement. When an indicator system (sheep erythrocytes coated with rabbit antibody) is then added, haemolysis takes place if there is free or unfixed complement.

Observe:

Absence of haemolysis: indicatingcomplement fixation due to initial reaction of antibody in the patient's serum with the original antigen— a positive result.

Presence of haemolysis: due to persistence of complement—a negative result.

4. Immunofluorescence

Usually indirect immunofluorescence in which smears containing the organism (antigen) are exposed to patient's serum followed by anti-human globulin labelled with a fluorescent dye. Serum antibody that has reacted with antigen attaches the labelled anti-human globulin: when viewed by ultra-violet microscopy the antigen fluoresces.

5. a. Radioimmunoassay (RIA)
b. Enzyme-linked immunosorbent assay (ELISA).
Both techniques, like immunofluorescence, depend on tagging anti-human globulin with a) a radio-isotope or b) an enzyme. Antibody in patient's serum is detected either by bound radioactivity or by a colour change produced by the bound enzyme on addition of a suitable substrate.

6

Sterilisation and disinfection

Doctors must know how to render articles safe from the risk of transmitting infection. This does not always require *sterility* which means the complete destruction of all organisms including spores. The degree of bacterial killing which is necessary depends on the circumstances. For example, pasteurised milk is far from sterile but is safe because any human pathogens of bovine origin that it might contain have been killed.

STERILISATION

Involves rendering an article sterile.

Bacteria in the vegetative—or non-sporing—state are readily killed by heat, e.g. at 56°C for 30 min or at 100°C for a few seconds.

Spores are survival mechanisms possessed by certain bacteria (in the field of medical microbiology all are members of the genera *Bacillus* and *Clostridium*) and—not surprisingly—are very resistant to heat and other inactivating agents: they are destroyed only by longer and more intensive application of the various methods used to kill bacteria.

Sterilisation is therefore complex, difficult and often costly: it is dependent on the following factors:

1. Knowledge of the killing or death curves of representative bacteria and spores when exposed to the inactivation process. Spores vary in the degree of their resistance to heat and other agents: each process should be tested against spores of appropriate resistance (*Bacillus stearothermophilus* for heat, *B. subtilis var. globigii* for ethylene oxide, *B. pumilus* for ionising radiation).
2. The penetrating ability of the inactivating agent—important in the case of surgical dressings or pads: steam penetrates much more effectively than dry heat.
3. The ability of the article to withstand the sterilisation process. Surgical fabrics tolerate steam but are destroyed by dry heat; plastics are heat-sensitive.

4. The effect of organic matter such as blood, faeces, dust, etc. which enhances the survival of bacteria and spores and interferes with the sterilisation process. Articles must be clean before they are sterilised.

Safety margin: sterilisation (and disinfection) must, in practice, always have a safety margin over and above the minimal treatment required for killing of bacteria and spores: this is arbitrarily fixed at 25 per cent and is to allow for possible inadequacies in the sterilisation process or for the presence of a few exceptionally resistant organisms or spores.

The various methods of sterilisation are shown in Table 6.1.

Table 6.1　Methods of sterilisation

Method	Equipment	Use
Moist heat (steam under pressure), e.g. at 121°C for 15 min or 134°C for 3 min	Autoclave	Surgical dressings, instruments, almost any article or fluid which is not heat-sensitive
Dry heat 160°C for 1 h	Hot air oven	Glassware, powders, ointments
Ethylene oxide (gaseous)	Special chamber	Plastic and rubber goods; certain sophisticated instruments; poor penetration therefore articles must be clean
Gamma irradiation	Carefully shielded cobalt-60 source. Expensive, special plant required. Used commercially	Plastic goods Orthopaedic prostheses
Subatmospheric steam at 80°C with formaldehyde vapour*	Special chamber	Heat-sensitive equipment, e.g. endoscopes
Filtration through cellulose membranes	Filtration apparatus	Heat-sensitive fluids (will not remove viruses)

*Subatmospheric steam without formaldehyde is often used to render equipment safe by killing vegetative bacteria.

Autoclaves

Steam is a very efficient sterilising agent because when it condenses to form water it:

1. Liberates latent heat which participates in bacterial killing.
2. Contracts in volume enhancing penetration.

Protein coagulates (and bacteria are killed) more readily in the presence of moisture.

When water is heated within a closed vessel the temperature at which it boils, and that of the steam it forms, rises above 100°C. Thus *steam temperature* rises with increase in atmospheric pressure: e.g. at 15 lb per in² the temperature of steam is 121°C, at 30 lb, 134°C: conversely at subatmospheric pressure the temperature of steam is below 100°C. The time of exposure necessary for sterilisation diminishes with rising temperature, e.g. at 134°C only 3 min is necessary.

The simplest autoclave is a domestic pressure cooker. It suffers two defects: it is impossible (i) to dry the load, (ii) to expel all the air because air is denser than steam.

An *autoclave* consists of a double-walled or jacketed chamber: steam circulates within the jacket and is supplied under pressure to the closed inner chamber in which the goods for sterilisation are placed (Fig. 6.1).

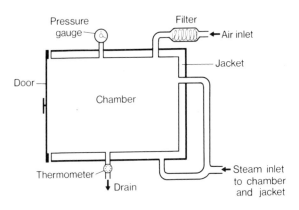

Fig. 6.1 Diagram of an autoclave (simplified).

Drying is accomplished at the end of the cycle by evacuating the steam from the chamber which contains the load while steam continues to circulate in the jacket so that the chamber remains hot. Moisture in the load evaporates and drying is assisted by sucking warmed filtered air into and through the chamber.

Expulsion of air is necessary because
1. The temperature of an air-steam mixture at a given pressure is lower than that of pure steam (Table 6.2).
2. Air pockets interfere with steam penetration.

Table 6.2 Autoclave pressures and temperatures

Chamber pressure	Chamber temperature of:		Usual method of air removal	Type of autoclave
	Pure steam	50:50 air-steam mixture		
15 lb	121°C	112°C	Gravity displacement by steam	Downward displacement
30 lb	134°C	128°C	Use of efficient pump	High vacuum

Downward-displacement autoclaves: obsolescent and relatively inefficient; a long cycle time is unavoidable. Although a preliminary vacuum may remove around half of the air, the remainder is displaced by admitting steam at the top of the chamber and allowing the heavier air to gravitate to the bottom where it is discharged. Only when this is complete can the sterilisation holding period commence: its duration depends on the time taken to penetrate the load and this is dictated by its composition and method of packing.

High-vacuum autoclaves: modern, fast and reliable equipment. Before the admission of steam, 98 per cent of the air present in the chamber is rapidly removed by an electric pump and sterilisation can proceed without delay. The cycle is short: all loads are penetrated efficiently.

The steam temperature recorded by the thermometer placed in the drain is used to indicate that the chamber is free from air e.g. if temperature is 121 °C and chamber pressure 15 lb, air evacuation is complete (See Table 6.2).

Centralised sterilisation facilities using high-vacuum autoclaves are available to most hospitals today.
1. *Central Sterile Supply Department (CSSD):* in which packs of sterilised dressings, equipment, etc. are prepared and then transported to the various wards, clinics and theatres: most—sometimes all—the sterilisation in the hospital is carried out in autoclaves in the CSSD by specially trained and supervised staff.
2. *Theatre Sterile Supply Unit (TSSU):* usually a subdepartment of CSSD specialising in theatre equipment, e.g. pre-set trays of surgical instruments.

Other sterilisation methods

The following are used:

Dry heat: by microbiology laboratories for glassware.

Ethylene oxide gas, gamma irradiation: largely by commercial suppliers for plastic goods.

Filtration: by pharmaceutical firms for the sterilisation of drugs for injection.

Sterile disposable plastic articles

Syringes, catheters, tubing, infusion bags etc. are now supplied as pre-sterilised disposable articles. The ready availability and comparative cheapness of these has revolutionised medical practice both in hospitals and general practice: they have also greatly improved safety since amateur attempts at sterilisation (especially of syringes) are no longer necessary.

Boilers

The small electrically heated boiler is a familiar piece of equipment in most doctors' and dentists' surgeries. The maximum temperature is, of course, 100 °C and will therefore not kill all spores. Nevertheless it is perfectly satisfactory for rendering safe from the risk of infection metal instruments such as specula, tongue depressors, forceps, etc. It is not suitable for syringes.

DISINFECTANTS AND ANTISEPTICS

Disinfectants and antiseptics are chemicals which inactivate vegetative bacteria but are rarely capable of killing spores. Most bacteriologists regard them as unreliable since their efficacy depends on exposure for lengthy periods of time and can be reduced or abolished by the presence of blood, faeces or other organic matter. Nevertheless they are extremely widely used in medicine.

Disinfectants: crude corrosive chemicals that are destructive to skin and mucous membranes; generally more active antimicrobial agents than *antiseptics* which can be applied to body surfaces.

Table 6.3 lists the most common agents: proprietary names are shown in brackets but this list includes only the preparations most familiar to the authors; many other equally effective proprietary-name brands are available.

Table 6.3 Disinfectants and antiseptics

Disinfectant	Antibacterial spectrum	Inactivation by organic matter	Comments
Phenolics			
1. Clear fluids* (Stericol, Clearsol, Hycolin)	Wide: but not spores	–	Cheap but irritant to skin
2. Chloroxylenol (Dettol)	Gram-positive bacteria	+	non-irritant but not very effective, wide domestic use
3. Hexochlorophane (Gamophen soap, Ster-Zac)	Gram-positive bacteria	+	Often incorporated in soap or detergent solution, effect is cumulative
Hypochlorites (Chloros, Diversol Bx, Milton, Eusol)	Wide, including viruses	+	Corrosive to metal, kills some spores
Povidone-iodine (Betadine, Disadine)	Wide	–	Hypersensitivity may be a problem
Formaldehyde	Wide, including viruses	–	Some effect on spores; used as gas or solution; irritant to eyes, respiratory tract
Glutaraldehyde (Cidex, Asep)	Wide, including viruses	–	Kills spores — slowly; penetration poor
Quaternary ammonium compounds, e.g. cetrimide (Cetavlon)	Gram-positive bacteria	+	Bacteriostatic: have detergent properties but inactivated by soap and man-made materials
70% alcohol. (ethyl or isopropyl)	Wide	–	Penetration poor: often used in combination with iodine or chlorhexidine
Chlorhexidine (Hibitane)	Gram-positive bacteria	+	Used in combination with cetrimide (Savlon)

* Clear fluids have almost completely replaced the crude black (e.g. Jeyes fluid) and white (e.g. Izal) coal tar preparations.

Bedpans

Safe disposal of faeces is a major problem in hospitals where a considerable proportion of the patients are confined to bed. Two main methods are now used:

1. Disposable papier-mâché bedpans: placed in a special apparatus which destroys the pan and flushes away the contents.
2. Stainless steel bedpans: locked into a disposal unit which empties them and flushes them out with steam.

The principal procedures for disinfection in hospital are shown in Table 6.4

Table 6.4 Hospital disinfection procedures

Article	Disinfectant or Antiseptic
Floors, walls*	Clear phenolic fluids— as 1% solution—2% if contaminated with pus or faeces
Surfaces, e.g. trays, locker tops, food preparation areas	Hypochlorite or 70% alcohol
Skin 1. Surgeons' hands	Chlorhexidine in either detergent or alcohol, povidone-iodine
2. Patient's skin at operation	Chlorhexidine in alcohol, iodine in alcohol, povidone-iodine
3. Patient's skin before venepuncture	70% alcohol
4. Cleansing of wounds, burns, pressure sores, ulcers	Aqueous solutions of chlorhexidine and cetrimide either alone or in combination: hypochlorites (Milton, Eusol)
Anaesthetic equipment, endoscopes	Gluteraldehyde Subatmospheric steam
Thermometers	Wipe with 70% alcohol; keep dry for individual patient

* A high standard of general domestic cleanliness is essential and often all that is required; the use of disinfectants may be restricted to dealing with infected spillages e.g. pus.
Note: Iodines are the only antiseptics that can kill spores on skin.

PUBLIC HEALTH ASPECTS—SAFE MILK AND WATER SUPPLIES

Pasteurisation of milk

One of the most successful applications of bacteriology: milk is raised to a temperature of either 63-66°C for 30 min or—in the flash method—to 72°C for 15 sec. Milk so treated is not sterile, as anyone who keeps it for a few days at room temperature will discover: but it is safe from contamination with viable *Mycobacterium tuberculosis*, brucellae, campylobacter, *Coxiella burneti* and other pathogenic vegetative bacteria.

Pasteurised milk is not spoiled in taste or appearance but the cream layer is slightly reduced. Almost all milk in Britain is now pasteurised and milk-borne episodes of infection have become rare.

Treatment of water

The natural habitat of some saprophytic bacteria is water and many soil bacteria gain access to water during heavy rain.

In addition, microorganisms of faecal origin from the human and animal intestine may find their way into water supplies: it is these that cause outbreaks of water-borne infection and the diseases spread in this way include enteric fever, dysentery, cholera, hepatitis A and gastroenteritis (sometimes without a recognised causal microorganism).

The number of bacteria in clean water in a good catchment area decreases rapidly on *storage*. Subsequently, before entering the piped supply, reservoir water is treated by
1. *Filtration* through sand supported on gravel and clinker
2. *Chlorination*

Medically important bacteria

7

Staphylococcus and micrococcus

Gram-positive cocci which are arranged in grape-like clusters.
Staphylococcus: pathogenic or commensal parasites.
Micrococcus: free-living saprophytes.

STAPHYLOCOCCUS

SPECIES:

1. *Staphylococcus aureus*—the main pathogen responsible for pyogenic injections.
2. *Staphylococcus epidermidis (Staphylococcus albus)*—a universal skin commensal.
3. *Staphylococcus saprophyticus* (previously classified as micrococcus subgroup 3)—similar to *Staph. epidermidis* but resistant to novobiocin.

Habitat: the body surfaces and, by dissemination, air and dust.

1. *Staph. aureus:* carried in the anterior nares (50-75 per cent of healthy people); less often in the skin (especially axilla and perineum) and mucous membranes (throat, gut). Originating from these reservoirs *Staph. aureus* is ubiquitous and always present in the hospital environment.
2. *Staph. epidermidis:* normally present in the resident skin flora; also found in gut, upper respiratory tract.

Laboratory characteristics

Morphology and staining. Gram-positive, non-sporing, non-motile cocci (diameter about 1 μm) characteristically arranged in clusters. (Plate 2).

Culture: grow well on ordinary media aerobically and, although less well, anaerobically; optimal temperature 37 °C.

Colonial appearance:

Staph. aureus: typically golden colonies but pigmentation varies from orange to white.

Staph. epidermidis: white colonies.

Pigment production with *Staph. aureus* is often poor after 24 h incubation at 37 °C on nutrient agar or blood agar; this important differential feature can be enhanced by leaving the plates in the light at room temperature. Pigmentation is encouraged by growth on special media, e.g. milk agar.

Selective media: staphylococci will tolerate sodium chloride in concentrations of 5 to 10 per cent. Salt-containing agar and broth are useful in isolating staphylococci from samples likely to contain large numbers of other bacteria.

Identification of Staph. aureus: since pigment production is an uncertain character, a number of tests have been developed to distinguish *Staph. aureus* from *Staph. epidermidis.*

1. *Coagulase test:* the conclusive test that distinguishes *Staph. aureus* (coagulase positive) from *Staph. epidermidis* (coagulase negative).

a. *Tube test.* Inoculate diluted human or rabbit plasma with the staphylococcus and incubate for 3-6 h at 37 °C.

Observe: clotting (gel formation) with coagulase—positive strains. (Fig. 7.1)

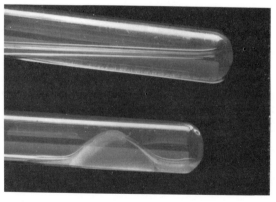

Fig. 7.1 Photograph of a positive tube coagulase test (bottom).

Mechanism: extracellular *free coagulase* produced on culture has thrombin-like activity and converts fibrinogen into fibrin.

b. *Slide test.* Emulsify the staphylococcus in a drop of saline and mix with a loopful of undiluted plasma.

Observe: rapid clumping of the suspension with coagulase-positive strains

Mechanism: bound coagulase (clumping factor) on the bacterial cell surface reacts directly with fibrinogen.

Although the reactions are due to two entirely different substances the vast majority of strains give either a positive or a negative result with both tests.

2. *Phosphatase test. Staph. aureus* produces a phosphatase; very few coagulase-negative strains are phosphatase-positive.
3. *Deoxyribonuclease test.* Coagulase-positive strains hydrolyse DNA; only a minority of coagulase-negative strains produce a DNAase.
4. *Mannitol fermentation.* Almost all *Staph. aureus* strains form acid from mannitol—a property possessed by few coagulase-negative strains. Mannitol-salt agar, containing an indicator to demonstrate mannitol fermentation, is a useful selective medium for identifying presumptive *Staph. aureus* in mixed cultures.

Typing:

Bacteriophage typing: strains of *Staph. aureus* can be distinguished by the pattern of their susceptibility to a set of 24 bacteriophages (phages); most strains are lysed by more than one phage and several hundred different phage types can be recognised. The phages are in three groups (I, II and III) to which the *Staph. aureus* strains can be assigned; these phage groups correspond broadly to three major serological groups identified antigenically.

Toxins:

Staph. aureus forms a large number of extracellular toxins and

Table 7.1 Toxins and toxic components produced by *Staphylococcus aureus*

Toxin	Activity
Haemolysins *a, ß, γ* and *o*	Cytolytic, lyse erythrocytes of various animal species
Coagulase	Clots plasma
Fibrinolysin	Digests fibrin
Leucocidin	Kills leucocytes
Hyaluronidase	Breaks down hyaluronic acid
DNAase	Hydrolyses DNA
Lipase	Lipolytic (produces opacity in egg-yolk medium)
Protein A	Antiphagocytic
Epidermolytic toxins A and B	Epidermal splitting and exfoliation
Enterotoxin(s)	Causes vomiting and diarrhoea
Toxic shock syndrome toxin-1	shock, rash, desquamation

enzymes. Not all strains produce the whole range listed in Table 7.1; most of the products probably play a role in pathogenicity.

Staph. epidermidis produces few if any of the toxins elaborated by *Staph. aureus.*

Pathogenicity

Staph. aureus, the main pathogenic species, is an important pyogenic organism: below is a list of some of the diseases it causes.

1. *Superficial infections:* pustules, boils, carbuncles, abscesses, impetigo, sycosis barbae, conjunctivitis, wound infections (including post-operative sepsis).
2. *Skin exfoliation:* toxic epidermal necrolysis (Ritter-Lyell's disease).
3. *Deep infections:* septicaemia, endocarditis, pyaemia, osteomyelitis, pneumonia.
4. *Toxic food poisoning.*

Staph. epidermidis is of much lower pathogenicity and rarely causes infections in healthy people: below are some of the diseases with which it is associated.

1. *Infective endocarditis* (especially with prosthetic heart valves).
2. *Infection of Spitz Holter valves* (in hydrocephalus).
3. *Infection of cannulae* (intra-vascular or intra-peritoneal)

Staph. saprophyticus is an important cause of urinary tract infection in sexually active women.

Antibiotic sensitivity

Staph. aureus shows exceptional ability to appear in multiply-resistant form—especially in hospitals. Sensitivity testing is therefore essential in the choice of an appropriate drug for therapy.

Antibiotics active against *Staph. aureus* are:

Penicillin (50 per cent of domiciliary and 80 per cent or more of hospital strains are now resistant)

Flucloxacillin (stable to β-lactamase produced by penicillin-resistant strains and therefore effective against them)

Erythromycin

Lincomycin and Clindamycin

Fusidic acid

Vancomycin

Cephalosporins

Penicillin resistance is due to production of β-lactamase which breaks down the antibiotic: the β-lactamase is plasmid-coded and transferred by transduction via bacteriophage.

In 1983-84, new strains of *multi-resistant Staph. aureus* (i.e. resistant to penicillin, flucloxacillin, tetracycline, erythromycin and sometimes also to gentamicin and fusidic acid) appeared in British hospitals. The infections caused were no more severe than those due to other strains but treatment of the occasional serious case of infection has had to rely largely on vancomycin.

Staph. epidermidis is often resistant to penicillin; most strains are sensitive to at least some of the other antistaphylococcal antibiotics. Vancomycin is particularly useful.

MICROCOCCI

Micrococci are often similar to and difficult to separate from coagulase-negative staphylococci. Many strains are strict aerobes and a useful differential test is their inability to form acid from glucose under anaerobic conditions.

Colonies are usually white but those of some species are brightly pigmented yellow, orange or pink.

Micrococci have little pathogenic potential.

ANAEROBIC CLUSTERING GRAM-POSITIVE COCCI

These are best considered with the other anaerobic cocci (see Chapter 19)—classification into 'anaerobic staphylococci' and 'anaerobic streptococci' is no longer considered justifiable.

8

Streptococcus and pneumococcus

Streptococci include pneumococci (also known as *Streptococcus pneumoniae*); however, pneumococci can be considered separately as they form a distinct species with its own properties and characteristics.

STREPTOCOCCI

Classification: is extremely complex and still incomplete. An important basis for classification is the type of haemolysis produced around colonies growing on blood agar (Table 8.1).

Table 8.1 Classification of streptococci based on haemolysis

Haemolysis	Appearance	Designation	Streptococcal class
Complete	Colourless clear, sharply defined zone	ß	Pyogenic streptococci
Partial	Greenish discolouration	α	Viridans streptococci
None	No change	δ or non-haemolytic	Enterococci

Note. This classification is not absolute and variable haemolysis is produced by some strains within the groups.

Pyogenic streptococci

This class includes the most pathogenic human species: the main pathogen is *Streptococcus pyogenes*. Pyogenic streptococci have polysaccharide Lancefield group antigens in their cell-wall: *Strep. pyogenes* is Lancefield group A; other pyogenic streptococci belong to Lancefield groups B, C, G, R and S.

Enterococci

All enterococci have the same glycerol-teichoic acid group antigen of Lancefield group D located between the cell wall and membrane; they may also possess the group G antigen; their normal habitat is the gut.

Viridans streptococci

A heterogeneous group, sometimes called indifferent or other streptococci: strains may possess one of a variety of Lancefield group antigens (A, C, E, F, G, H, K, M or O) or none at all.

GENERAL PROPERTIES OF STREPTOCOCCI

Morphology and staining: Gram-positive spherical or oval cocci in pairs or chains (Plate 3): 0.7-0.9 μm diameter.

Culture: grow well on blood agar: enrichment of media with blood, serum or glucose may be necessary.

Selective media containing aminoglycosides or 1:500 000 crystal violet inhibit other bacteria in a mixed culture but permit growth of streptococci.

Lactic acid is the major fermentative product: it accumulates in cultures and rapidly terminates growth. Culture in buffered glucose medium (e.g. Todd-Hewitt broth) increases the yield of organisms.

Aerobic: but grow well, sometimes better, *anaerobically*: growth enhanced by 10 per cent carbon dioxide.

Colonial morphology: usually small, except for enterococci.

Haemolysis: see Table 8.1.

Biochemical Reactions: all give a negative catalase reaction. Streptococci can be characterised by their biochemical activities and a commercial API system which combines 20 tests is useful in identification particularly of species belonging to the viridans group.

Serology: identifies *Lancefield groups*: 20 are recognised designated A-H and K-V. The antigens that define the groups are either polysaccharide or teichoic acid.

Method: classically a precipitin reaction with antigen extracted by acid treatment; now carried out by slide agglutination using commercially-prepared kits.

Bacitracin. *Strep. pyogenes* is generally more sensitive to bacitracin than other haemolytic streptococci. This property is the basis of a disc-diffusion sensitivity test for the *presumptive* identification of Lancefield group A streptococci. However, some streptococci of other groups including strains belonging to groups C and G may be bacitracin-sensitive by this test and this gives rise to confusion.

The main medically important Lancefield groups with the species usually responsible for human disease are as follows:

Group A: *Strep. pyogenes*
Group B: *Strep. agalactiae*
Group C: *Strep. equisimilis*
Group D: *Strep. faecalis*
Group G: no species recognised
Groups R and S: *Strep. suis*

PYOGENIC STREPTOCOCCI

LANCEFIELD GROUP A STREPTOCOCCI

STREPTOCOCCUS PYOGENES

The most pathogenic member of the genus: produces a large number of powerful enzymes and toxins.

Habitat: present as a commensal in the nasopharynx of a variable proportion of healthy adults and children.

Laboratory characteristics

Culture: blood agar with small, typically matt or dry colonies surrounded by β haemolysis.

Capsule: many strains produce a hyaluronic acid capsule during the logarithmic phase of growth and develop mucoid colonies on blood agar. Also formed by some strains in Lancefield Group C. May be associated with resistance to phagocytosis.

Toxins: the following extracellular products have been characterized.

1. *Streptokinase*
   a. a protease which lyses fibrin by catalysing conversion of plasminogen to plasmin
   b. Two immunologically-different types produced—A and B
   c. Also produced by some strains of streptococci in Lancefield groups B, C, G and F
2. *Hyaluronidase*
   a. Attacks hyaluronic acid—the cement of connective tissue—causing increased permeability
   b. Antibodies to this enzyme are produced after infection

3. *DNAases (deoxyribonucleases)*
   a. Four immunologically distinct types—A, B, C and D; B enzyme is the most common
   b. Antibodies to DNAase—especially B enzyme—are demonstrable in patients after most infections with *Strep. pyogenes*
   c. Also produced by some streptococci in Lancefield groups B, C, G and L.
4. *NADase (nicotinamide adenine dinucleotidase)*
   a. Kills leucocytes
   b. Antibody formed after infection
   c. Also formed by some steptococci in Lancefield groups C and G
5. *Haemolysins:* two streptolysins—or toxins which lyse erythrocytes—are produced; their characteristics are shown in Table 8.2.

Table 8.2 Streptolysins of *Streptococcus pyogenes*

Haemolysin	Stability to oxygen	Synthesised anaerobically	Antigenic
Streptolysin O	−	+	+
Streptolysin S	+	−	−

+ = yes
− = no

6. *Erythrogenic toxin*
   a. Produced as a result of the presence of a lysogenic phage in the streptococci
   b. Three immunologically distinct toxins—A, B and C
   c. Responsible for the characteristic erythematous rash in scarlet fever
   d. *Dick test.* Injection intradermally produces localised area of erythema within 8-16 h in non-immune people; presence of antibody to the toxin results in neutralisation of the toxin and no reaction is produced
7. Other enzymes are produced and include:
   a. Leucocidin
   b. Protease
   c. Amylase

Note. All these enzymes and toxins probably contribute to the invasiveness and pathogenicity of *Strep. pyogenes*.

Serotypes
Strep. pyogenes (Lancefield group A): can be subdivided into Griffith types depending on three surface protein antigens:

M: type-specific, i.e. there is a distinct M antigen for each type or strain; only found in virulent or pathogenic strains; M antigens impede phagocytosis, antibody to them enhances phagocytosis; 65 distinct M serotypes have been identified.

R: not related to virulence; fewer antigens and the same R antigen can be found in several different M types. Also found in a few strains in Lancefield groups B, C and G

T: there are also T antigens each one of which may be found on several different M types; used in conjunction with M typing for identifying different types of *Strep. pyogenes*

Pathogenicity

Streptococcus pyogenes causes:
  Tonsillitis and pharyngitis
  Peritonsillar abscess (quinsy)
  Scarlet fever
  Otitis media
  Mastoiditis and sinusitis
  Wound infections; may lead to cellulitis and lymphangitis
  Impetigo
  Erysipelas; an acute lymphangitis of the skin
  Puerperal sepsis
Post-streptococcal complications:
  Rheumatic fever
  Glomerulonephritis
  Erythema nodosum

Antibiotic sensitivity

Penicillin: drug of choice.

  In patients hypersensitive to penicillin: erythromycin

  Antibiotic resistance: is rare—penicillin resistance is unknown but resistance to tetracycline is not uncommon.

LANCEFIELD GROUP B STREPTOCOCCI

Group B contains only one species: *Strep. agalactiae*—increasingly recognised as an important human pathogen. Three main types with subdivisions are recognised: human strains (mainly Type I) are distinct from animal strains.

  Habitat: commensal of female genital tract; this may be secondary to ano-rectal carriage: common in animals—especially cattle—in which it causes bovine mastitis.

Culture: grows on ordinary and bile-containing media (eg. MacConkey).

Colonies: usually β but may be α or non-haemolytic: typically produces red or orange pigment when incubated anaerobically on serum- and starch-containing media (e.g. Islam's medium).

Identification: Lancefield grouping.

Pathogenicity: an important pathogen in neonates causing meningitis and septicaemia: also associated with septic abortion, puerperal or gynaecological sepsis.

Antibiotic sensitivity: penicillin, erythromycin.

LANCEFIELD GROUP C STREPTOCOCCI

Rarely cause human disease.

Culture: blood agar; colonies often large and mucoid due to hyaluronic acid capsules. Usually ß-haemolytic.

Toxins: elaborate a number of extra cellular substances antigenically similar to those of *Strep. pyogenes.*

Identification: Lancefield grouping.

Biotypes: four species can be distinguished by biochemical and other tests:

Strep. equisimilis	most common species isolated from humans: produces streptokinase
Strep. equi	rarely
Strep. dysgalactiae	found
Strep. zooepidemicus	in humans

Note. Some strains of *Strep. milleri* (see under viridans streptococci) have group C antigen.

Pathogenicity. Low pathogenicity for man: occasionally cause of tonsillitis especially in closed institutional communities; rarely endocarditis, glomerulonephritis.

Antibiotic sensitivity: penicillin.

LANCEFIELD GROUP G STREPTOCOCCI

A heterogeneous group: no species have so far been recognised: share a number of characteristics with Lancefield groups A and C

Habitat: human throat

Culture: two colonial forms both producing β-haemolysis

1. Large colony: strains resemble *Strep. pyogenes*: pathogenicity low, tonsillitis, rarely endocarditis.
2. Small colony: strains are occasionally found in the human throat, possibly gut: low pathogenicity—rarely abdominal sepsis, tonsillitis.

Note. Some strains of *Strep. milleri* possess Group G antigen.

LANCEFIELD GROUPS R and S STREPTOCOCCI

One species *Strep. suis* is recognised
Habitat: pigs
Culture: on blood agar colonies usually *β*—but occasionally *α*-haemolytic.
Identification: Lancefield grouping differentiates 2 types.
Strep. suis (Serotype 1)—S antigen
Strep. suis (Serotype 2)—R antigen
Both types may also possess the D antigen.
Pathogenicity: Both types cause serious infections in pigs. Serotype 2 strains cause zoonotic infection, notably meningitis, in farmers, abattoir workers and butchers.

ENTEROCOCCI

LANCEFIELD GROUP D STREPTOCOCCI

This group includes the enterococci, a common and important cause of human disease.

Morphology: oval cocci, usually in pairs, do not readily chain.

Culture: grow on ordinary and bile-containing media: heat-resistant and able to grow at 45 °C; also able to grow in the presence of 6.5 per cent sodium chloride.

Table 8.3 Group D streptococci

Species	Habitat	Disease
Strep. faecalis *Strep. faecium* *Strep. durans* (true enterococci)	Human and animal gut	Urinary, abdominal wound infection: rarely endocarditis
Strep. bovis (sometimes classified as a viridans streptococcus)	Animal gut especially ruminants; sometimes human gut	Endocarditis

Colonies: large, whiteish, haemolysis variable but generally non-haemolytic; lactose-fermenting colonies on MacConkey agar (pink) and on CLED agar (yellow).

Identification: colonial and cultural characteristics; antibiotic sensitivity pattern; demonstration of Lancefield group D antigen.

Species: the medically important species are listed in Table 8.3.

Antibiotic sensitivity: enterococci are sensitive to ampicillin, moderately resistant to penicillin and resistant to the cephalosporins.

VIRIDANS OR INDIFFERENT STREPTOCOCCI

An ill-defined group which typically show α haemolysis on blood agar: but haemolysis is variable and some strains are non-haemolytic: most human strains are commensals of the upper respiratory tract and of low pathogenicity.

Clinical laboratories usually do not differentiate species but simply report *Strep. viridans:* species identification within the viridans group depends on the results of a range of biochemical tests: species do not possess a characterising Lancefield group antigen.

The principal species are listed in Table 8.4.

Table 8.4   Viridans (indifferent) streptococci

Species	Haemolysis on blood agar	Lancefield group antigens	Habitat	Disease
Strep. mitior (*Strep. mitis*)	α (β or none)	O,K,M,Q or none		Endocarditis
Strep. sanguis	α (β or none)	H,K or none	Human oropharynx	Endocarditis
Strep. mutans	None	E or none		Endocarditis dental caries
Strep. salivarius	None	K or none		Rarely, endocarditis
Strep. milleri	None (α or β)	A,C,F,G or none	Human oropharynx, gut, vagina	Deep abdominal, liver, lung, brain, abscesses

Note: Strep. milleri is now recognised as an important cause of sepsis and is sometimes classified with the pyogenic streptococci

(   ) = less common reactions

STREPTOCOCCUS PNEUMONIAE

Probably the most common and nowadays the most important pathogen amongst the streptococci:

Also known as pneumococcus or *Diplococcus pneumoniae.*

Habitat: normal commensal of the upper respiratory tract.

Laboratory characteristics

Morphology: lanceolate diplococci (Plate 4) arranged longitudinally in pairs with the pointed ends outwards: sometimes forms short chains: normally capsulated with carbohydrate antigenic capsule which is correlated with virulence.

Culture: blood agar, sometimes broth enriched with serum or glucose.

Colonies: α-haemolytic, typically 'draughtsman', i.e. with sunken centre due to spontaneous autolysis of older organisms: young colonies may resemble dew drops due to large capsules before autolysis.

Identify: differentiate from viridans streptococci by:-
1. *Optochin sensitivity:* viridans streptococci are resistant.
2. *Bile solubility:* addition of bile to broth cultures lyses pneumococci but not viridans streptococci.
3. *Inulin fermentation:* viridans streptococci do not ferment inulin.

Antigenic structure:

1. *Capsule:* contains the polysaccharide carbohydrate antigen: type specific: 83 capsular types are recognised.
   Identified by capsule swelling—quellung reaction—observed microscopically when pneumococci are mixed with specific antisera.
2. *C substance:* a cell-wall-associated antigen common to all pneumococci: consists of choline teichoic acid.
3. *Protein M antigen:* resembles but unrelated to M antigens of *Strep. pyogenes.* Not associated with virulence.

Virulence correlates with the presence of a capsule probably because this prevents or inhibits phagocytosis.

Transformation: the transfer of DNA—and some of the genetic markers for which it codes—from one bacterial strain to another was first demonstrated in pneumococci by Griffiths in 1928.

Pneumococcal types: not all types are equally common and the majority of human infections are associated with a limited number of types.

Common infecting types: these are included in the polyvalent vaccine used in prophylactic immunisation (see p. 354 for list of types).

Invasive types: some of the infecting types are particularly liable to cause serious infections such as pneumonia, septicaemia and meningitis e.g. types 1, 3, 6 and 8.

Pathogenicity

Pneumococci are important pathogens and cause a considerable amount of both morbidity and mortality today despite their sensitivity to penicillin.

Lobar pneumonia
Acute exacerbation of chronic bronchitis (often with H. *influenzae*)
Meningitis
Otitis media
Sinusitis
Conjunctivitis
Septicaemia (especially in splenectomized patients)

Antibiotic sensitivity

Sensitive to penicillin, erythromycin.

9

Enterobacteria

Gram-negative bacilli which belong to the tribe Enterobacteriaceae. Often called 'coliform bacilli', or coliforms. Intestinal parasites of man and animals, many are human pathogens.

Table 9.1 lists the main medically important species.

Table 9.1 Medically important enterobacteria

Genus	Species	Principal diseases
Escherichia	*Esch. coli*	Wound and urinary infection; gastroenteritis in children
Shigella	*Sh. dysenteriae* *Sh. flexneri* *Sh. boydii* *Sh. sonnei*	Dysentery
Salmonella	*S. typhi* *S. paratyphi* A,B,C	Enteric fever
	S. typhimurium Many other serotypes	Food-poisoning
Klebsiella	*K. pneumoniae* *K. oxytoca*	
Morganella	*Morg. morgani*	
Proteus	*Pr. mirabilis* *Pr. vulgaris*	Urinary infections, other forms of sepsis
Providencia	*Prov. stuartii* *Prov. rettgeri* *Prov. alcalifaciens*	
Yersinia	*Y. pestis*	Plague
	Y. pseudotuberculosis *Y. enterocolitica*	Septicaemia, enteritis, mesenteric adenitis
Enterobacter	*Ent. cloacae* *Ent. aerogenes*	
Serratia Citrobacter	*Serr. marcescens* *C. freundii*	Generally of lower pathogenicity

Laboratory characteristics

Morphology and staining. Gram-negative bacilli 2–3.0 × 0.6 μm: non-motile or motile by peritrichous flagella (Fig. 9.1): non sporing.

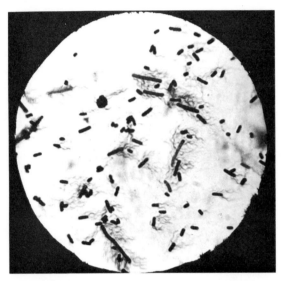

Fig. 9.1 Flagella. Peritrichous flagella demonstrated by a silver-impregnation staining method (magnified × 2000).

Culture. Grow well on ordinary media, e.g. blood agar, MacConkey agar, CLED agar; aerobic and facultatively anaerobic; wide temperature range.

Identification:

1. *Lactose fermentation* on indicator media assists in initial identification: e.g. MacConkey agar contains lactose and a pH indicator so that lactose-fermenting colonies can be identified by their pink colour. On CLED medium the colour change of lactose-fermentation is from blue-green to yellow.
2. *Biochemical tests:* some are characteristic of all enterobacteria, e.g.
   a. Reduce nitrate to nitrite
   b. Ferment glucose with acid (and sometimes gas) production
   c. Oxidase negative
   Others are required to identify different species: now usually done by test kits which allow rapid and early inoculation of 10 (AP1 10E) on 20 (AP1 20E) biochemical tests.

3. *Serological tests* based on identification of somatic and flagellar antigens are used for the final identification of *Salmonella* and *Shigella species.*
Strain identification within a species can also be done by
a. Bacteriophage typing
b. Bacteriocine typing.

Antigenic structure is complex, with considerable sharing of antigens both between species and within a species: antisera prepared in animals, and usually requiring numerous absorptions to reduce cross-reacting antibodies, are used to identify an organism on the basis of the specific antigens it possesses. A diagram of the sites of the main antigens in enterobacteria is shown in Figure 9.2.

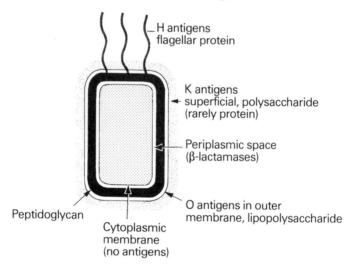

Fig. 9.2 Antigenic structure of Gram-negative bacteria.

Endotoxins are *lipopolysaccharides* consisting of various sugars and lipid A: present in the cell wall of Gram-negative bacilli (see Chapter 2), they are liberated when the bacterial cells lyse and are responsible for many pathological effects during human infection with the organisms.

Exotoxins are *proteins* liberated extracellularly from the intact bacterium by a few species of enterobacteria, e.g. *Shigella dysenteriae* produces a neurotoxin, toxigenic strains of *Escherichia coli* produce enterotoxins.

Antibiotic sensitivity

Antibiotic sensitivity is difficult to predict for any individual strains due to the frequency with which enterobacteria acquire plasmids which may spread to other strains: plasmids usually contain genes which code for resistance to several antibiotics.

Antibiotics used against enterobacteria are listed below:
Ampicillin/amoxycillin and mezlocillin
Aminoglycosides
Trimethoprim
Chloramphenicol
Cephalosporins (particularly of the second and third generation)
Urinary antimicrobial drugs: in addition to those listed above, some other antibacterial drugs can be used, but only for urinary tract infections, e.g.
Nitrofurantoin
Nalidixic acid

ESCHERICHIA COLI

Habitat: a normal inhabitant of the human and animal intestine.
Isolation: grows well as large colonies after overnight incubation.
Identification. Most strains are motile: some strains are capsulate. Usually ferments lactose (hence produces colonies pink on MacConkey agar, yellow on CLED agar). Indole is produced. Further biochemical tests are required for formal identification.
Typing:
1. Determination of O antigens.
2. Colicine typing: colicines are bacteriocines produced by *Esch. coli.*
Pathogenicity:
1. Urinary tract infection
2. *Diarrhoea.* There are three types of diarrhoea-producing strains:
   a. Enteropathogenic (EPEC)—infantile diarrhoea
   b. Enterotoxigenic (ETEC)—tourist diarrhoea
   c. Enteroinvasive (EIEC)
3. Wound infections—especially after lower intestinal tract surgery (the flora of these infections often includes anaerobes).
4. Peritonitis.
5. Biliary tract infections.
6. Septicaemia.
7. Neonatal meningitis.

SHIGELLA

Habitat: intestinal parasites of man

Isolation: shigellae grow well on routine media. They do not ferment lactose and so produce colourless colonies on MacConkey agar and DCA (a medium with lactose which also contains desoxycholate citrate to inhibit growth of *Esch. coli*).

Shigella sonnei is an important exception, it ferments lactose slowly with production of pale pink colonies.

Enrichment culture of faeces in selenite F broth, with subsequent subculture onto MacConkey agar may improve isolation.

Identification. There are four species and numerous serotypes:

Sh. dysenteriae, 10 serotypes;

Sh. boydii, 14 serotypes;

Sh. flexneri, six serotypes;

Sh. sonnei, one serotype.

1. *Motility:* shigellae are non-motile.
2. *Biochemical tests:* shigellae produce acid but not gas from carbohydrates.
3. *Determination of O antigens.*

   Typing: 'colicine' typing is available for *Sh. sonnei.*

   Pathogenicity:

Sh. dysenteriae, which produces protein exotoxins, causes the most severe illness, 'Shiga dysentery'. Dysentery due to other shigellae tends to be milder: *Sh. sonnei* is the cause of most dysentery in Britain.

SALMONELLA

Habitat. Animal gut: predominantly animal pathogens which can also cause disease in man. Foodstuffs from animal sources are important vehicles in the transmission of infection. *Salmonella typhi* and *S. paratyphi* differ from the other species in that man is the only natural host.

Isolation: grow well on ordinary media.

From faeces, culture on:

1. *MacConkey agar.* ⎤ Observe for pale (non-lactose
2. *Desoxycholate citrate agar* ⎦ fermenting) colonies.
3. *Wilson and Blair's bismuth sulphite agar:* observe for black metallic colonies (due to hydrogen sulphide production).
4. *Selenite F and tetrathionate broths* for enrichment: then subculture to MacConkey agar.

Identification:

1. *Motility:* salmonellae are motile
2. *Biochemical tests:* salmonellae generally produce acid and gas from carbohydrates—except for *S. typhi* which does not produce gas.
3. *Serology:* by identification of antigens:
   O: somatic
   H: flagellar
   Vi: a surface antigen possessed by a few species notably *S. typhi*
   More than 1500 serotypes ('species') are recognised: sharing of O and H antigens is common and identification is complex depending on detection of several antigens (see Table 9.2).

Table 9.2 Selected salmonellae showing antigenic profiles (Kauffmann-White scheme)

Serotype ('species')	Group	O Antigens	H Antigens Phase 1	Phase 2
S. paratyphi A	A	1.2.12.	a	—
S. paratyphi B	B	1.4.5.12.	b	1.2
S. agona	B	4.12.	f.g.s.	—
S. typhimurium	B	1.4.5.12.	i	1.2
*S. paratyphi C**	C	6.7.	c	1.5
*S. typhi**	D	9.12.	d	—

*Also possess Vi antigen

Phase variation: a single strain can possess two different sets of H antigens at different times; both sets must be analysed for identification.

Typing: Bacteriophage typing of particular serotypes can be carried out e.g. Vi-phages of *S. typhi.*

Pathogenicity:

1. *Enteric fever* is due to *S typhi, S. paratyphi A, B, C*
2. Most other salmonella serotypes (*S. typhimurium* is the most common) cause gastroenteritis or food-poisoning; however, some particular types (e.g. *S. dublin, S. cholerae-suis*) have a tendency to cause septicaemia.
3. Rarely, salmonellae cause osteomyelitis, septic arthritis, and other purulent lesions.

KLEBSIELLA

Habitat. Human and animal intestine. Some strains are saprophytes in soil, water and vegetation. They survive well in moist environments in hospitals.

Isolation. Grow well, producing colonies which are often but not always large and mucoid (due to the possession of a prominent capsule) on blood agar, CLED agar and MacConkey agar.

Identification: Not motile. Biochemical tests: lactose is fermented. Nomenclature is confused, and there is disagreement as to the names and number of species in the genus; *Klebsiella pneumoniae, K. oxytoca* are terms commonly used.

Typing:

1. Antigenic analysis of the capsular polysaccharides which comprise the large capsules typical of this genus. More than 80 serotypes recognised.
2. Klebicine (bacteriocine) typing.

Pathogenicity:

1. Urinary tract infection.
2. Septicaemia.
3. Meningitis (especially in neonates).
4. Rarely abscesses, endocarditis and other lesions. Sometimes associated with chronic nasal and oropharyngeal sepsis.
5. Friedlander's pneumonia.

PROTEUS

Habitat. Human and animal intestine. Some strains are saprophytes and are found in soil and water.

Isolation: grow well on routine media: a swarming type of growth which may cover the whole plate is produced on ordinary media by *Proteus mirabilis* and *Pr. vulgaris.* Proteus forms colourless colonies on MacConkey agar and blue-green colonies on CLED agar.

Identification: Swarming growth allows presumptive identification of *Pr. mirabilis* or *Pr. vulgaris.* Biochemical tests permit formal identification. Lactose is not fermented. *Pr. mirabilis* is indole-negative. Proteus species produce a potent urease.

Typing:

1. Serotyping.
2. Bacteriocine typing.
3. 'Dienes' phenomenon—inhibition of swarming by dissimilar strains of *Pr. mirabilis.*

Pathogenicity:

Pr. mirabilis is the most frequently isolated species.

1. Urinary tract infection: urinary urea is 'split' by the bacterial urease to produce ammonium salts (alkaline pH); association with urinary calculi.
2. Often isolated from mixed flora of wounds, burns, pressure sores, chronic discharging ears—generally a low-grade pathogen in such circumstances but occasionally these sites provide a portal of entry for septicaemia or brain abscess (ear infections).
3. Septicaemia.

YERSINIA

Habitat. Yersinia are animal parasites which sometimes—although rarely—cause disease in man.

Isolation: from blood or tissue (when samples are available); difficult from mixed flora of faeces.

Colonies are often smaller than those of other enterobacteria, and prolonged incubation of cultures is therefore necessary.

Identification. Generally small (1.5 × 0.5 μm) bacilli which may show 'bipolar' staining (i.e. darker at both ends of the bacilli). *Yersinia pestis* is non-motile; *Y. pseudotuberculosis* and *Y. enterocolitica* are motile at 22°C. Biochemical tests permit formal identification, the reactions being more reproducible at 22-27°C than the 37°C temperature used for other enterobacteria.

Identification of *Y. pestis* is simplified by fluorescent-antibody staining and by testing for susceptibility to specific bacteriophage.

Pathogenicity:

1. *Plague: Y. pestis* (formerly called *Pasteurella pestis*) causes bubonic and pneumonic plague ('The Black Death').
2. *Y. pseudotuberculosis* and *Y. enterocolitica* cause
   a. A septicaemic illness similar to typhoid fever
   b. Enteritis
   c. Mesenteric adenitis, sometimes associated with terminal ileitis. Clinically, this closely mimics appendicitis

ENTEROBACTER; SERRATIA; PROVIDENCIA; MORGANELLA; CITROBACTER

These members of the enterobacteria can conveniently be considered together.

Habitat. Human and animal intestinal parasites, but some strains are saprophytes. Moist environments in hospitals may be important reservoirs.

Isolation. They grow well on routine media.

Identification: biochemical tests.

The most frequently isolated species are:

Enterobacter cloacae, Ent. aerogenes, Serratia marcescens, Providencia rettgeri, Prov. stuartii, Citrobacter freundii, Morganella morgani.

Pigment: Some strains of *Serr. marcescens* produce characteristic magenta-coloured colonies but hospital strains are non-pigmented.

Pathogenicity

1. Urinary tract infections (particularly chronic, complicated infections).
2. Wounds, skin lesions and respiratory infections in hospitalised patients.
3. Septicaemia.

Some have been responsible for outbreaks of infection in intensive care areas, burns units and other special units; they are often resistant to many antibiotics.

10

Pseudomonas, miscellaneous aerobic Gram-negative bacilli

PSEUDOMONAS

Gram-negative motile aerobic bacilli having very simple growth requirements: widely distributed in water, soil and sewage: the genus includes pathogens for animals and plants.

PSEUDOMONAS AERUGINOSA (PSEUDOMONAS PYOCYANEA)

Habitat: human and animal gastrointestinal tract, water, soil. Moist environments are important reservoirs in hospitals and *Pseudomonas aeruginosa* is able to survive and multiply in certain aqueous antiseptics and other fluids.

Laboratory characteristics

Morphology and staining: Gram-negative bacilli (1.5–3.0 μm × 0.5 μm), motile (polar flagellate), non-sporing, non-capsulate.

Culture: a strict aerobe; grows readily on all routine media over a wide temperature range (5 °C to 42 °C).

Broth cultures: show uniform turbidity and a surface pellicle— evidence of its aerobic growth requirement.

Plate cultures: some colonies are 'coliform-like' but others typically large and irregular; they have a fluorescent, greenish appearance due to production of *pyocyanin* (blue-green) and *fluorescein* (yellow) pigments which also diffuse into the surrounding medium. The cultures develop a characteristic 'fruity' odour. Strains isolated from sputum of patients with cystic fibrosis often give rise to large mucoid colonies due to formation of extracellular polysaccharide slime: such colonies may fail to produce pyocyanin.

Selective media:

1. *MacConkey agar:* yellowish non-lactose fermenting colonies.
2. *Cetrimide agar:* this medium inhibits many other organisms but allows *Ps. aeruginosa,* which resists the action of quaternary ammonium compounds, to grow.

Identify by:

Colonial morphology.

Biochemical tests: acid production in a range of carbohydrates which are oxidised not 'fermented': oxidase-positive (this is a valuable differential test); liquifies gelatin.

Examination of cultures under ultraviolet light: colonies display intense greenish fluorescence—useful to detect fluorescent pseudomonads in mixed cultures.

Ability to grow at 42°C: this distinguishes *Ps. aeruginosa* from other pseudomonads that produce fluorescein.

Typing:

Pyocine (bacteriocine) typing: a method that can be carried out in routine diagnostic laboratories. The sensitivity of eight indicator strains of *Ps. aeruginosa* to the pyocines produced by the test strain is noted and from the pattern of growth inhibition about 40 types can be recognised. Unfortunately lacks discrimination because a few types are very prevalent.

Phage typing: considerable difficulty has been encountered in obtaining reproducable results.

Antigenic structure: a typing scheme exists based on approximately 20 heat-resistant somatic ('O') antigens and heat-labile flagellar ('H') antigens. Available only in reference centres it is said to be both reproducable and discriminatory.

Pathogenicity

An important cause of hospital-acquired infections. Most severe infections occur in patients with serious underlying conditions (e.g. burns, malignancy), or as a result of therapeutic procedures (e.g. indwelling urinary tract catheters, mechanical ventilatory support). Previous antibiotic therapy also favours infection with *Ps. aeruginosa.*

1. Urinary tract infection:
   a. chronic, complicated infections
   b. association with indwelling catheter.
2. Burns.
3. Septicaemia—'ecthyma gangrenosum' skin lesions may be present.
4. Wound infections.
5. Infected skin lesions, e.g. pressure sores, varicose ulcers.
6. Chronic otitis media and externa.
7. Lower respiratory tract infections
   a. in cystic fibrosis
   b. in patients on ventilators.
8. Eye infections, secondary to trauma or surgery. Loss of sight may result.

Vaccine

Sixteen serotypes of *Ps. aeruginosa* are now internationally recognised and a polyvalent vaccine made from the cell-surface components of representative strains has been developed. This vaccine, although not yet commercially available, stimulates active immunity in man (and protective human immunoglobulin for passive immunisation has been produced from the serum of vaccinated volunteers). It may be of value in the prevention of pseudomonas infections in susceptible patients: the efficacy of vaccination (active and passive) has been demonstrated in patients with severe burns in field trials conducted in India.

Antibiotic sensitivity

Resistant to most usual antibiotics. The major antipseudomonal drugs are:

Aminoglycosides

Certain *ß*-lactams: penicillins (carbenicillin, ticarcillin, azlocillin, piperacillin); cephalosporins (cefsulodin, ceftazadime).

Polymyxins B and E (colistin).

Note. A laboratory report of *Ps. aeruginosa*, particularly if isolated from a superficial site, is not an indication for prescription of toxic and expensive antibiotics. A careful clinical and bacteriological assessment of the likely relevance of the information in the individual patient should always be made.

OTHER PSEUDOMONAS SPECIES

PSEUDOMONAS FLUORESCENS AND PSEUDOMONAS PUTIDA

These fluorescent pseudomonas are similar to *Ps. aeruginosa* and are often regarded as biotypes of that species. They differ in their failure to produce pyocyanine, in some biochemical reactions and in their inability to grow at 42 °C. Some strains will not grow even at 37 °C. However many strains grow at 4 °C. This property has enabled *Ps. fluorescens* and *Ps. putida* to multiply in stored contaminated blood: such blood, if transfused, causes severe reactions.

Other non-fluorescent species of *Pseudomonas* which lead a primarily saprophytic existence are recognized. Compared to *Ps. aeruginosa*, their ability to cause human infection is limited. However, such organisms (e.g. *Ps. cepacia, Ps. maltophilia*) do occasionally give rise to serious human infection, particularly when host defences are impaired.

PSEUDOMONAS MALLEI AND PSEUDOMONAS PSEUDOMALLEI

These antigenically related bacteria cause similar diseases. They are now recognised to be non-fluorescent pseudomonads but in the past they were classified in genera that have been discarded, e.g. *Pfeifferella, Loefflerella, Malleomyces.*

Ps. pseudomallei grows readily but *Ps. mallei*, unlike other pseudomonads, grows poorly on ordinary laboratory media, is non-motile and is a strict animal parasite.

Ps. pseudomallei (Whitmore's bacillus) causes melioidosis, a disease of animals and humans, endemic in South-East Asia. Most human cases are asymptomatic. Pulmonary consolidation, skin lesions and a fatal septicaemia may occur. The organism is a saprophyte of certain soils and waters and there may be a large animal reservoir locally.

Ps. mallei causes glanders in horses. Rarely, human infections occur, acquired from animals or from laboratory work with the organism.

OTHER AEROBIC GRAM-NEGATIVE BACILLI

ACINETOBACTER

Gram-negative cocco-bacilli (microscopic morphology may be confused with neisseria). Strictly aerobic, non-motile, oxidase-negative.

Habitat: widely distributed in nature. Human skin carriage may be an important cause of dissemination in hospitals.

Isolation: grows well on routine media.

Identification: some strains form acid from sugars oxidatively (*Aci. calcoaceticus*) others are asaccharolytic (*Aci. lwoffi*).

Pathogenicity: generally a low-grade pathogen: increasingly being isolated from hospitalised patients, particularly from those with burns and in intensive care units. May colonise intravascular cannulae.

Antibiotic sensitivity: variable: often resistant to many antibiotics, including those commonly used to treat other forms of serious hospital sepsis.

11

Vibrio and campylobacter

Actively motile, Gram-negative curved bacilli.

VIBRIO

Widespread in nature, mainly in water: one species, *Vibrio cholerae*, is the cause of *cholera*.

VIBRIO CHOLERAE (VIBRIO COMMA)

Habitat: water contaminated with faeces of patients or carriers.

Laboratory characteristics

Morphology and staining. Gram-negative slender bacilli ($2 \times 0.5\,\mu$m), sometimes comma-shaped with a pointed end; often arranged in pairs or short chains giving a spiral appearance. Actively motile by one long polar flagellum; non-capsulate, non-sporing.

Culture. Aerobe; grows readily on ordinary media as glistening colonies over a wide temperature range (optimum 37 °C). Growth is inhibited at acid pH (optimal pH for growth is alkaline from 8.0-8.2).

Enrichment medium. Alkaline peptone water (pH 8.4) promotes the rapid growth of *V. cholerae* from mixtures of other bacteria, e.g. in faecal samples.

Selective media:

1. TCBS medium—thiosulphate citrate bile sucrose agar pH 8.6
   Observe: for large yellow sucrose-fermenting colonies after 18-24 h incubation. Enterobacteria may grow but growth is inhibited and the colonies are small.
2. Monsur's medium—gelatin taurocholate tellurite agar pH 8.5-9.2
   Observe: for large colonies, grey at 24 h, with a black centre at 48 h, surrounded by a halo due to gelatin liquefaction.

Identify by slide agglutination of suspect colonies with specific antisera.

Biochemical reactions. Fermentation of sucrose and mannose (acid, no gas) but not arabinose is typical of *V. cholerae* and a distinguishing feature from other vibrios.

Cholera red reaction: add a few drops of sulphuric acid to a 24 h peptone-water culture.

Observe: for immediate formation of red-pink colour (this reaction is given by many vibrios and is not specific for *V. cholerae*).

Mechanism: production of indole and nitrites in the peptone water results in formation of the coloured substance nitroso-indole.

Biotypes. Two biotypes of *V. cholerae:* (1) classical and (2) El Tor— can be differentiated (biotyping is the distinguishing of different bacterial strains within a species by various biological and biochemical reactions).

Antigenic structure:

H antigens are common to many vibrio species but of little value in identification.

O antigens enable the vibrios to be divided into six major subgroups. All strains of *V. cholerae* possess a distinctive 0 antigen and belong to subgroup I with subdivision into three serotypes, *Ogawa, Inaba* and *Hikojima*; however, antigenic structure (and, therefore, serotype) may change within the human gut. Any serotype can be either classical or El Tor biotype.

Vibrios dificient in the O antigen of subgroup I are classified as non-cholera or non-agglutinable vibrios (NCV or NAG).

Phage typing. Five phage types of Classical *V. cholerae* and six phage types of *V. cholerae* El Tor have been described. Phage typing is of only limited value in epidemiological studies.

Toxins. Endotoxins (cell-wall lipopolysaccharide) and exotoxins are recognised. The enterotoxin is an exotoxin which stimulates persistent and excessive secretion of isotonic fluid by the intestinal mucosa.

Pathogenicity

Vibrio cholerae is the cause of cholera in man. In the acute disease vibrios are present in enormous numbers—about 10^8 per ml of faeces; some animals are susceptible to laboratory challenge and can serve as models for the study of the disease.

Viability. Readily killed by heat and drying; dies in polluted waters but may survive in clean stagnant water (especially if alkaline) or sea water for 1 to 2 weeks. El Tor biotype is more resistant to adverse conditions than classical *V. cholerae*.

Antibiotic sensitivity

Sensitive to a wide range of antibiotics including tetracycline and chloramphenicol.

VIBRIO PARAHAEMOLYTICUS

V. parahaemolyticus is a halophilic marine vibrio isolated from shellfish particularly in countries with warm coastal waters, e.g. South-East Asia.

It causes an acute gastroenteritis in which vibrios are excreted in large numbers in the stools. Faecal samples plated on TCBS agar yield large blue-green colonies typical of *V. parahaemolyticus*.

CAMPYLOBACTER

Strictly microaerophilic vibrios have been placed in a new genus— *Campylobacter*. The main type species is *Campylobacter fetus*.

Subspecies: several are recognised but their nomenclature is controversial and confused. The main human pathogen is: *C. fetus* subspecies *jejuni* (sometimes known as *C. jejuni*)

Other subspecies: most do not cause human disease although they are important pathogens in domestic animals. One— *C. fetus* subspec. *intestinalis* is a rare cause of human infection.

Campylobacter fetus subspec. *intestinalis* is found in the intestine of cattle, sheep and birds. It causes abortion in sheep. It grows well at 25 °C but not at 43 °C.

Laboratory characteristics of *C. fetus* subsp. *jejuni*

Morphology and staining. Small, Gram-negative, curved or spiral rods (Plate 5); highly motile by a single flagellum at one or both poles.

Culture. Microaerophilic; grow best in an atmosphere containing a mixture of 7 per cent oxygen, 10-15 per cent carbon dioxide with the remainder an inert gas, usually nitrogen or hydrogen.

Growth takes place at 37 °C but the optimal temperature is 43 °C.

Media. Grow readily on simple media.

Selective medium (necessary for isolation from faeces and other samples containing numerous other bacteria): lysed blood agar with vancomycin, polymyxin and trimethoprim.

Incubate: for 24-48h at 43°C under microaerophilic conditions.

Observe: for effuse colonies that look like spreading fluid droplets.

Identify b: Gram-film appearance, motility, growth temperature requiremen' (25°C-, 37°C+, 43°C++) and oxidase test (campylobacters are oxidase-positive).

Pathogenicity

Responsible for febrile diarrhoeal illness in man: occasionally cause septicaemia.

Antibiotic sensitivity

Sensitive to macrolides (e.g. erythromycin) and aminoglycoside antibiotics, partially sensitive to ampicillin, resistant to penicillin.

AEROMONAS: PLESIOMONAS

Gram-negative bacilli; aerobes facultative anaerobes: motile; oxidase-positive.

Aeromonas hydrophila and *Plesiomonas shigelloides* are the medically important species but infections are rare and usually in patients with some other serious disease: occasionally isolated from blood, CSF and exudates. Both organisms may be isolated from human faeces but the significance of this finding is uncertain. They are *not* accepted as an established cause of diarrhoeal disease.

12

Parvobacteria

Parvobacteria (*parvus*=small) are a heterogeneous group which contains several important human pathogens causing a wide variety of different diseases. Parvobacteria are small, Gram-negative bacilli which generally require enriched media for isolation and culture. Parvobacteria contain the following genera:

Haemophilus
Brucella
Bordetella
Pasteurella
Francisella
Actinobacillus.

HAEMOPHILUS

Habitat: mainly the respiratory tract: often part of the normal flora but may also cause respiratory disease.

Laboratory characteristics

Morphology and staining. Small Gram-negative coccobacilli (Plate 4), non-sporing, non-motile.

Culture: requires enriched media such as blood or chocolate agar: optimum temperature around 37 °C: most species grow poorly in the absence of oxygen and growth is enhanced in an atmosphere with added CO_2; enriched media are necessary because *Haemophilus species* need one or both of two growth factors:

X factor—haematin
V factor—diphosphopyridine nucleotide.

Requirement for growth factors can help to differentiate species (Table 12.1).

Table 12.1 Growth factors for *Haemophilus species*

Factor required	Species
X and V	*H. influenzae, H. aegyptius,* *H. haemolyticus*
X	*H. ducreyi*
V	*H. parainfluenzae, H. parahaemolyticus*

Pathogenicity

The diseases caused by *Haemophilus* species are listed in Table 12.2.

Table 12.2 Pathogenicity of *Haemophilus species*

Species	Disease
H. influenzae	Exacerbations of chronic bronchitis Meningitis Epiglottitis Sinusitis, otitis media
H. aegyptius	Conjunctivitis
H. ducreyi	Chancroid
H. parainfluenzae *H. haemolyticus* *H. parahaemolyticus*	Commensals of the upper respiratory tract; rarely cause disease

HAEMOPHILUS INFLUENZAE

The main pathogenic species.

Habitat: the upper respiratory tract: most strains found in the normal flora are non-capsulated.

Laboratory characteristics

Morphology: small Gram-negative coccobacillus; a minority of strains are capsulated.

Culture: on chocolate or blood agar: a streak of *Staphylococcus aureus* across the plate produces V factor and enlarges the size of adjacent colonies of *H. influenzae*—satellitism (Fig. 12.1).

Colonial morphology: small translucent, non-haemolytic colonies: capsulated strains form larger iridescent colonies.

Selective medium: the addition of bacitracin to chocolate agar inhibits the growth of many Gram-positive bacteria (e.g. viridans streptococci) found in the upper respiratory tract and use of this medium facilitates the isolation of *H. influenzae* from sputum specimens.

Identify: by testing on nutrient agar for requirements for X and V growth factors.

Antigenic structure

Serotypes. Six are recognized—on the basis of capsular polysaccharide antigens—Pittman types a, b, c, d, e, f. Type b is the main cause of haemophilus meningitis and acute epiglottitis.

Pathogenicity

Non-capsulated strains are mainly responsible for exacerbations of chronic bronchitis and bronchiectasis.

Capsulated strains (predominately type b) can cause various infections mainly in children aged from 2 months to 3 years:

Meningitis ⎤ these infections often
Acute epiglottitis ⎥ accompanied by
Osteomyelitis ⎥ septicaemia.
Arthritis ⎦

Fig 12.1 Satellitism. A blood agar plate showing enhancement of the growth of colonies of *Haemophilus influenzae* next to the streak of *Staphylococcus aureas* which supplies V factor.

Antibiotic sensitivity

Sensitive to:
Ampicillin
Cotrimoxazole
Tetracycline
Erythromycin
Chloramphenicol (reserve for meningitis and epiglottitis)
'Second and third generation' cephalosporins.

Penicillin: strains are generally not fully sensitive to penicillin; in a small proportion, resistance to penicillin and ampicillin is due to β-lactamase production.

HAEMOPHILUS DUCREYI

The cause of the sexually transmitted disease *chancroid,* or soft sore.

Laboratory characteristics

Morphology: slender Gram-negative ovoid bacilli, slightly larger than *H. influenzae,* bacteria *en masse* have configuration of 'shoals of fish'.

Culture: on 20–30 per cent rabbit blood agar or other special enriched medium: incubate at 33 °C for 3-5 days with added moisture and CO_2.

Colonial morphology: small grey glistening colonies surrounded by a zone of haemolysis.

Growth factors: only X factor is required.

Antibiotic sensitivity: sulphonamide (the drug of choice in treatment).

BRUCELLA

Predominantly infect domestic animals from which infection may be transmitted to man.

SPECIES

Three main species *Brucella melitensis, Br. abortus* and *Br. suis* each with a number of biotypes are usually recognised: some of the subtypes are associated with a particular geographical location.

Habitat: chronic infection in domestic animals (see Table 12.3).

Table 12.3 Animal hosts and geographical distribution of *Brucella species*

Strain	Animal host	Geographical distribution
B. melitensis	Goats, sheep	Mediterranean countries
B. abortus	Cattle	Worldwide including UK
B. suis	Pigs	Denmark and USA

Laboratory characteristics

Morphology and staining: short, slender, pleomorphic, Gram-negative bacilli: non-mobile, non-sporing.

Culture: in enriched medium such as liver infusion broth or agar; colonies develop after several days incubation at 37 °C in aerobic conditions; CO_2 is required for the growth of *Br. abortus.*

Identification of the different species is done by a variety of biochemical and serological tests and inhibition by dyes: this is illustrated in Table 12.4.

Table 12.4 Differentiation of *Brucella species*

	Br. melitensis	Br. abortus	Br. suis
Growth requirement for CO_2	–	+	–
Production of H_2S	–	+	+
Growth in presence of			
Basic fuchsin	+	+	–
Thionine	+	–	+
Methyl violet	+	+	–
Agglutination by monospecific antibody to			
Br. abortus	–	+	+
Br. melitensis	+	–	–

Antigenic structure: the three species share two antigens, A and M, but these are present in different proportions. Typical melitensis strains contain an excess of M antigen whereas typical abortus and suis strains contain an excess of A antigen. Monospecific antisera can be prepared and these are of use in identification.

Pathogenicity

The cause of undulant fever or brucellosis—a chronic, debilitating febrile illness usually without any localising signs: the bacteria persist intracellularly and are therefore difficult to eradicate by antibiotic therapy.

Antibiotic sensitivity

Tetracycline combined with streptomycin is the treatment of choice.

BORDETELLA

The only important member of the species is *Bordetella pertussis*, the cause of whooping cough. *Bord. parapertussis* causes a milder form of whooping cough which is uncommon in Britain.

BORDETELLA PERTUSSIS

Habitat. The human respiratory tract usually associated with acute disease.

Laboratory characteristics

Morphology and staining: short, sometimes oval, Gram-negative bacilli: freshly isolated strains may be capsulated.

Culture: special enriched medium is required for primary isolation: the most widely used media are charcoal blood agar and Bordet-Gengou medium usually made selective by the addition of penicillin.

Bordet-Gengou medium contains 30 per cent blood, potato extract, glycerol, agar.

Colonial morphology: colonies like 'split pearls' appear after three or more days of incubation in a moist aerobic atmosphere at 35 °C.

Identification is confirmed serologically by slide agglutination with a polyvalent antiserum.

Antigenic structure: surface antigens (agglutinogens) designated 1 to 6 are recognised: all freshly isolated strains possess agglutinogen 1.

Serotypes: there are three main serotypes based on the presence of these surface antigens: type 1, 2; type 1, 2, 3; type 1, 3.

Pathogenicity

The cause of whooping cough: a disease mainly seen in pre-school children and especially severe in those under 1 year of age. Affects the lower respiratory tract causing bronchospasm and the characteristic *paroxysmal cough.*

Antibiotic sensitivity

Erythromycin.

PASTEURELLA

PASTEURELLA MULTOCIDA

The main pathogenic member of the genus, also known as *P. septica*.

Habitat: respiratory tract of many animals, notably dogs.

Morphology: small, sometimes capsulated, ovoid Gram-negative bacilli often showing bipolar staining.

Culture: on blood agar (does not grow on MacConkey agar—a differentiating feature from enterobacteria).

Pathogenicity: an important animal pathogen: in man may cause septic wounds after dog or cat bites.

Antibiotic sensitivity: gentamicin, tetracyclines—the drugs of choice.

FRANCISELLA

FRANCISELLA TULARENSIS

The cause of tularaemia.

Habitat: rodents and small mammals.

Morphology: pleomorphic, small Gram-negative coccobacilli often showing bipolar staining.

Culture: on blood agar enriched with cysteine and glucose.

Pathogenicity: tularaemia, a plague-like disease of rodents, is contracted by contact with animal hosts or their products: seen in USA and occasionally in Europe.

Antibiotic sensitivity: streptomycin is the drug of choice: tetracyclines are also generally effective.

ACTINOBACILLUS

ACTINOBACILLUS SPECIES

Facultatively anaerobic, non-branching Gram-negative coccobacilli that grow on blood agar. The type species *Actinobacillus lignieresi* is responsible for actinobacillosis in cattle and sheep, a disease which resembles actinomycosis in man. *Actinobacillus actinomycetemcomitans* is sometimes found along with *Actimomyces species* in human actinomycosis.

13

Corynebacterium and related bacteria

CORYNEBACTERIUM

Gram-positive bacilli with a characteristic morphology: non-sporing; non-capsulate; non-motile.

Some strains are widely distributed in soil and plants. There are several human and animal species that are important pathogens and commensals.

CORYNEBACTERIUM DIPHTHERIAE

Habitat: the throat and nose of man.

Laboratory characteristics

Morphology and staining: pleomorphic Gram-positive rods (3 × 0.3 μm) or clubs which divide by 'snapping fission' so that adjacent cells lie at different angles to each other forming V, L and W shapes—so-called *Chinese-character* arrangement; adjacent cells may also lie parallel to one another in *palisades*.

Some strains stain irregularly due to intracellular deposition of polymerised phosphate forming the metachromatic or volutin granules characteristic of, but not exclusive to, *C. diphtheriae*. The granules, usually two or three per cell, show up with special stains—bluish black by Albert's method, deep blue with Neisser's methylene blue.

Culture: an aerobe and facultative anaerobe; optimum temperature 37°C. Does not grow well on ordinary agar and media containing blood or serum are required. Selective media are necessary for isolation from clinical specimens.

Selective media:

1. *Loeffler's serum medium: C. diphtheriae* grows rapidly, faster than

other upper respiratory tract bacteria present in clinical material: the morphology develops particularly well and smears made as soon as 8 hours after inoculation may show a typical appearance (Table 13.1).

2. *Blood tellurite agar* (e.g. Hoyle's or McLeod's medium); after 48 h incubation, corynebacteria produce characteristic grey-black colonies due to their ability to reduce potassium tellurite to tellurium.

Three colonial types of *C. diphtheriae* are recognised, *gravis*, *intermedius* and *mitis*; they were so named from the severity of the clinical disease they cause, but the correlation with colonial morphology is not good (Table 13.1). Other corynebacteria also grow on tellurite media to form colonies that can be confused with *C. diphtheriae*.

Table 13.1 Characteristics of *Corynebacterium diphtheriae* strains

Corynebacterium diphtheriae	Appearance in film from Loeffler's medium	Colonial type on tellurite medium
Gravis strains	Club-shaped, few metachromatic granules	Flat, grey with raised centre and irregular edge. Radial striations develop to form a 'daisy-head'
Intermedius strains	Short irregularly staining rods without metachromatic granules but in Chinese character arrangement	Small, smooth colonies of uniform size; grey-black with paler periphery
Mitis strains	Classic morphology with numerous granules and typical arrangement	Medium-sized, circular convex, glistening and black

Identify: by biochemical tests and demonstration of toxin production. Other corynebacteria may mimic *C. diphtheriae* in films and on culture. Some isolates of *C. diphtheriae*, especially mitis strains, are not toxigenic and therefore non-virulent.

Biochemical reactions: acid production from a range of carbohydrates and other biochemical tests are used to differentiate *C. diphtheriae* from other corynebacteria. Gravis (but not intermedius or mitis) strains ferment starch and glycogen.

Typing. Serotyping by agglutination tests, phage typing and bacteriocin typing have all been used to subdivide strains of *C. diphtheriae* for epidemiological studies.

Toxin production: is responsible for virulence; can be demonstrated by guinea pig inoculation or by a gel precipitation test.

1. *Guinea pig inoculation:* inject suspension of the isolated strain of *C. diphtheriae* into two guinea pigs, one protected with diphtheria antitoxin (Table 13.2).

Table 13.2 Test for toxin production by *C. diphtheriae*

Result	Unprotected animal	Antitoxin protected animal
Toxigenic strain	Death in 2-3 days	Survival
Non-toxigenic strain	Survival	Survival

2. *Gel-precipitation (Elek) test.* A filter paper strip previously immersed in diphtheria antitoxin is incorporated into serum agar before it has set; the strain of *C. diphtheriae* under investigation is then streaked onto the agar at right angles to the filter paper strip. Incubate at 37 °C.

Observe after 24h and 48h for precipitation indicating toxin – antitoxin interaction (Fig. 13.1).

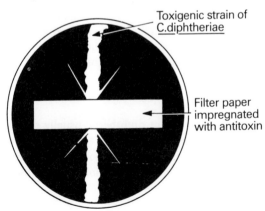

Fig. 13.1 Elek test for demonstration of toxin production by *Corynebacterium diphtheriae.* The toxin combines with antitoxin to produce antigen-antibody complexes which form visible lines of precipitation in the agar.

Diphtheria toxin. The exotoxin of *C. diphtheriae* is produced only by strains carrying a bacteriophage: its formation *in vitro* is stimulated in culture media with low iron content. The toxin interferes with protein biosynthesis in mammalian cells by splitting the molecule of NAD (nicotinic adenine dinucleotide)—an essential cofactor for the transferase involved in peptide bond formation by ribosomes. It acts locally on the mucous membranes of the respiratory tract to produce a grey adherent pseudomembrane consisting of fibrin, bacteria, epithelial and phagocytic cells: after absorption into the bloodstream, it acts systemically on the cells of the myocardium, the nervous system (only motor nerves are affected) and the adrenals.

The toxin can be rendered non-toxic but still antigenic by treatment with formaldehyde: the *toxoid* so formed is used in prophylactic immunisation.

Schick test: is a skin test to demonstrate immunity i.e. circulating diphtheria antitoxin—the result either of previous immunisation or infection (clinical or subclinical).

Method: intradermal injection of toxin into the anterior aspect of one forearm and heat-inactivated toxin into the other.

Object: to detect susceptibility to diphtheria toxin and hypersensitivity to the toxin or other more heat-stable proteins produced by *C. diphtheriae*. In children over 10 years of age and in adults, a preliminary Schick test should be carried out before immunisation to detect pseudo-reactors who may experience severe allergic responses to the vaccine (see Table 13.3).

Observe: for erythema at the injection site: reactions due to the toxin are slower to develop and longer lasting than those resulting from hypersensitivity. The arm should be examined at 1-2 days (e.g. 36 hours) and again at 5-7 days (e.g. 120 hours).

The reactions are summarized in Table 13.3.

Pathogenicity

C. diphtheriae is the cause of diphtheria: usually the mucous membranes of the upper respiratory tract are affected but sometimes, especially in tropical countries, skin lesions are produced: the serious systemic manifestations follow absorption of the exotoxin.

Antibiotic sensitivity

C. diphtheriae is sensitive to penicillin, erythromycin, the lincomycins and other antibiotics.

Table 13.3 The Schick test

Result	Test arm (toxin) 36h	120h	Control arm (inactivated toxin) 36h	120h	Interpretation	Diphtheria immunisation
Negative	—	—	—	—	Immune, not hypersensitive	Not required
Positive	±	+	—	—	Non-immune, not hypersensitive	Required
Negative and pseudo	+	—	+	—	Immune, hypersensitive	Not required
Positive* and pseudo 'combined'	+	+	+	—	Non-immune, hypersensitive	Contra-indicated

*Combined reactions are rare.

OTHER CORYNEBACTERIA

CORYNEBACTERIUM ULCERANS

May be responsible in man for diphtheria-like throat lesions but with little evidence of toxaemia.

Biochemical reactions distinguish it from *C. diphtheriae.*

Toxins: two are produced—one immunologically identical to the toxin of *C. diphtheriae*, the other related to the toxin of *C. ovis.*

ANIMAL PATHOGENS

A number of corynebacteria (*C. ovis, C. murium, C. pyogenes, C. equi, C. renale*) are important animal pathogens. The first two are morphologically similar to *C. diphtheriae.*

HUMAN COMMENSALS

There are many so-called *diphtheroid bacilli*, the most often isolated being *C. hofmannii* in the throat and *C. xerosis* in the conjunctiva.

Habitat: normally present in the skin (especially within sebaceous ducts) and mucous membranes.

Morphology and staining: less pleomorphic and more strongly

Gram-positive than *C. diphtheriae;* metachromatic granules are few or absent. Tend to be arranged in palisades with less pronounced Chinese lettering.

Culture: grow well on ordinary agar.

Pathogenicity: occasional opportunistic pathogens, e.g. causing endocarditis on prosthetic valves, infection in implanted artificial joints.

PROPIONIBACTERIUM

Gram-positive, non-sporing, anaerobic bacilli formerly classified as anaerobic corynebacteria. A differential feature from other similar anaerobes is that they produce propionic acid as a breakdown product of carbohydrate fermentation; this can be detected by gas-liquid chromatography.

Two main species are recognised, *Propionibacterium acnes* and *P. granulosum.*

Habitat: the human skin.

Pathogenicity: apparent association with the skin disease *acne vulgaris.*

ERYSIPELOTHRIX

ERYSIPELOTHRIX RHUSIOPATHIAE

A slender gram-positive non-motile bacillus, sometimes filamentous: related to the corynebacteria.

Habitat: healthy pigs but widely distributed in other animals and birds: found on the skin and scales of fish: *causes* swine erysipelas.

Pathogenicity: responsible in man for erysipeloid—a rare skin infection.

LISTERIA

LISTERIA MONOCYTOGENES

Morphologically similar to erysipelothrix but flagellated; feebly motile at 37°C but exhibits active tumbling motility at 25°C in young broth cultures.

Habitat: wild and domestic animals.

Culture: aerobic and facultatively anaerobic; optimal temperature 37°C but will survive and grow at 5°C.

Colonies on blood agar surrounded by a narrow zone of complete (β) haemolysis.

Pathogenicity: causes listeriosis in man.

14

Mycobacterium

Mycobacteria are widely distributed in nature: referred to as acid-fast because after mordanting in stain they resist decolorisation with strong acids. Although Gram-positive they stain poorly, if at all, by Gram's method. These staining properties are related to the complex composition of the cell wall which has a high lipid content. Acid fastness is the result of the formation of complexes between the dye and mycolic acid, one of the cell-wall lipids.

Species are divided into rapid growers which form colonies on a suitable medium in 2-3 days and slow growers which take 1-3 weeks. The main medically important mycobacteria are shown in Table 14.1 together with some of their properties.

Table 14.1 Medically important mycobacteria

Species	Habitat and source	Disease	Cultural characteristics on Lowenstein-Jensen medium
Myco. tuberculosis	Infected humans	Tuberculosis	Rough, dry yellow colonies: slow grower
Myco. bovis	Infected cattle	Tuberculosis	White, smooth colonies (inhibited by glycerol): slow grower
Myco. leprae	Infected humans	Leprosy	No growth
Atypical mycobacteria	Mainly soil, water, sometimes birds, animals	Pulmonary infection, cervical adenitis, skin ulcers	Often pigmented colonies: some species grow slowly others rapidly; may exhibit unusual temperature requirements

MYCOBACTERIUM TUBERCULOSIS

Laboratory characteristics

Morphology: slender, beaded bacilli: non-sporing.

Staining: does not stain by Gram's method. Ziehl-Neelsen stain must be used: smears are treated with concentrated carbol fuchsin, mordanted by heating and then decolorised with 20 per cent sulphuric acid and alcohol; the bacilli retain a bright red colour after this treatment and show up clearly against the background of counterstain (usually malachite green or methylene blue) (Plate 7). Nowadays usually detected under fluorescent microscopy with auramine staining.

Culture: obligate aerobe: does not grow on ordinary media: grows well—*eugonic growth*—on Lowenstein-Jensen medium (contains whole egg, asparagine, glycerol and—to inhibit contaminants—malachite green) usually after 2-3 weeks' incubation at 37 °C; cultures should be kept for 6-8 weeks before being discarded.

Pathogenicity

The cause of tuberculosis—a slowly progressive, chronic infection usually of the lungs but many other organs and tissues may become affected: the main source of infection is respiratory secretions from a patient with 'open' disease.

Pathogenic for guinea pigs, with production of typical progressive disease and death in 6-8 weeks; rabbits are much less susceptible.

MYCOBACTERIUM BOVIS

The main cause of infection in tubercular cattle: man becomes infected by ingestion of milk containing *Myco. bovis*.

Culture: grows relatively poorly—*dysgonic growth*—on Lowenstein-Jensen medium: growth much improved if glycerol replaced by pyruvic acid.

Pathogenicity: similar to human tubercle bacilli but particularly liable to infect children causing enlarged, casious cervical lymph nodes ('scrofula') and tuberculosis of the bones, joints and kidneys.

Pathogenic for both guinea pigs and rabbits with development of generalised disease.

Rare nowadays due to eradication of disease in cattle.

Antibiotic sensitivity

Both *Myco. tuberculosis* and *Myco. bovis* are sensitive to a wide range of drugs (listed below); because resistant variants arise readily, therapy should always be a combination of drugs.

Isoniazid
Rifampicin
Ethambutol
Streptomycin
Pyrazinamide
Para-aminosalicylic acid
Thioacetazone

MYCOBACTERIUM LEPRAE

The cause of leprosy—still a scourge in many parts of the Third World today.

Habitat: the organism is found only in cases of human infection.

Cultivation: in vivo by inoculation of the footpads of mice or of armadillos: the animals develop slow-growing granulomas at the site of injection. Does not grow *in vitro*.

Antibiotic sensitivity
Sensitive to:
Dapsone (a sulphone)
Rifampicin
Thiacetazone

ATYPICAL MYCOBACTERIA

A group of miscellaneous mycobacteria of low pathogenicity for man.

Culture: generally grow on Lowenstein-Jensen medium sometimes at lower (25°C) or higher (45°C) temperatures than normal. Several species produce pigmented growth in the dark (scotochromogens), some only after exposure to light (photochromogens) whereas others are non-pigmented.

Classification: controversial; the original Runyon groups I-IV now not often used.

Species. the principal species are shown in Table 14.2 together with the diseases they cause.

Table 14.2 Atypical mycobacteria

Species	Disease
Myco. avim/intracellulare	
Myco. chelonei	pulmonary
Myco. fortuitum	lymphaduopathy
Myco. marinum	abscesses
	swimming pool granuloma
Myco. ulcerans	tropical skin ulcers
Myco. kansasii	pulmonary
Myco. scrofulaceum	cervical adenitis
Myco. xenopi	pulmonary

Sources: many atypical mycobacteria are acquired from water notably—*Myco. marinum, Myco. ulcerans, Myco. xenopi* and *Myco. kansasii.* Other species, e.g. *Myco. avium/intracellulare* also infect birds and animals.

Pulmonary infection: in many cases, atypical mycobacteria are simply 'passengers' which accompany tuberculosis, and treatment— even when the atypical mycobacteria are resistant—usually results in cure.

Antibiotic sensitivity: variable, often resistant to several of the standard antituberculous drugs.

15

Actinomyces and nocardia

Actinomyces and nocardia are morphologically similar Gram-positive branching rods and filaments. Actinomyces are microaerophilic or anaerobic on primary isolation although some species grow in air after a few subcultures; nocardia are aerobic organisms.

ACTINOMYCES

Most actinomyces are soil organisms but some—and these are the potentially pathogenic species—are commensals of the mouth of man and animals (Table 15.1).

Table 15.1 Some *Actinomyces species*

Species	Host/habitat	Disease association
Actino. israelii *Actino. eriksoni*	Oropharynx and gut of man	Human actinomycosis
Actino. naeslundii *Actino. viscosus* *Actino. odontolyticus*	Oropharynx of man	Dental plaque and caries
Actino. bovis	Oropharynx of cattle	Lumpy jaw in cattle

Species are identified by colonial appearances (some are pigmented), ability to grow aerobically and biochemical tests.

ACTINOMYCES ISRAELII

Laboratory characteristics

Morphology and staining. Gram-positive bacteria which grow in filaments that readily break up into rods and may show branching. Non-motile, non-sporing, not acid-fast. In tissue, colonies develop to form diagnostic yellowish 'sulphur granules' which are visible to the naked eye and which are found in pus discharged through draining sinuses (Plate 9).

Culture:

Solid media: on blood or serum glucose agar incubated anaerobically at 37°C for 7 days or more; growth is enhanced by 5 per cent carbon dioxide.

Observe: for small, cream, adherent, nodular colonies.

'Shake' cultures in semi-solid glucose agar kept at 37°C for 5-10 days.

Observe: for maximal growth in a turbid band 10-15 mm below the surface where conditions are microaerophilic.

Isolation of this exacting microorganism from clinical material is difficult especially as the pus often contains other faster-growing bacteria. Presumptive diagnosis is made by the demonstration of typical Gram-positive branching filaments in a sulphur granule. Whenever possible, a washed, crushed sulphur granule should be cultured in preference to pus.

Pathogenicity

Actinomycosis is endogenous in origin and results in a chronic granulomatous infection with abscess formation: profuse pus discharges by draining through sinuses. Infection probably starts after local trauma, e.g. the extraction of carious teeth, appendicectomy. The typical sites of the disease are: cervico-facial—65 per cent of cases: abdominal (usually ileo-caecal)—20 per cent of cases; and, rarely, thoracic, affecting the lung.

Intrauterine contraceptive devices: are often colonized into *Actino israelii.* The significance of this is uncertain but, perhaps in association with other organisms, the organism may cause low-grade intrauterine infection.

Antibiotic sensitivity

Sensitive to penicillin, lincomycin, tetracycline, erythromycin.

NOCARDIA

Habitat: majority of species are soil saprophytes; a few are pathogenic to man.

Morphology and staining: similar to actinomyces but some species are weakly acid-fast.

Culture: slow-growing, aerobic organisms which require 5-14 days incubation on nutrient agar.

Observe: wrinkled, rosette to star-shaped colonies, initially white, then yellow and finally pink or red.

Antibiotic sensitivity: sensitive to sulphonamides and cotrimoxazole: resistant to practically all antibiotics. Treatment may have to be continued for six months.

Pathogenicity: cause chronic granulomatous suppurative infections.

NOCARDIA ASTEROIDES

Affects lungs, sometimes with secondary brain abscess.

Opportunistic infection: pulmonary nocardiosis has been described in renal transplant and other immunocompromised patients.

NOCARDIA MADURAE AND NOCARDIA BRASILIENSIS

Affect subcutaneous tissues: 'madura foot' or mycetoma is a tropical form of nocardiosis which affects the foot and produces chronic discharging sinuses.

STREPTOBACILLUS

Streptobacillus moniliformis is a slender branching filamentous bacterium with club-shaped ('moniliform') terminal swellings. Gram-negative but may be Gram-positive in young cultures. Requires enriched media for growth. Causes one form of rat-bite fever in man.

16

Neisseria and branhamella

NEISSERIA

The neisseriae are Gram-negative diplococci: the two pathogenic species *Neisseria gonorrhoeae* (the gonococcus) and *Neisseria meningitidis* (the meningococcus) have exacting growth requirements; the commensal neisseriae are easy to culture.

NEISSERIA GONORRHOEAE

Habitat: an obligate parasite of the human urogenital tract.

Laboratory characteristics

Morphology and staining. Gram-negative oval cocci 0.6-1.0 μm occur characteristically in pairs with adjacent sides flattened or concave ('bean-shaped') and long axes parallel. In purulent clinical material many of the diplococci are intracellular within a relatively small number of the polymorphs, the remainder extracellular in the exudate (Plate 10). Some pleomorphism and variation in the intensity of staining is common.

Culture: requires an enriched medium (usually a lysed-blood or chocolate agar) and incubation in a moist aerobic atmosphere containing 5-10 per cent CO_2. Optimal temperature 35-37°C.

Selective media: enriched agar media can be made selective by the addition of antibiotics which inhibit other bacteria but not *N. gonorrhoeae*, e.g. Thayer-Martin medium which contains vancomycin, colistin, nystatin and trimethoprim: some recent modifications contain lincomycin in place of vancomycin. Used for the isolation of *N. gonorrhoeae* from clinical material especially when a wide range of other bacteria will be present, e.g. specimens from vagina and rectum.

Observe: small grey glistening colonies after 24h incubation becoming larger, opaque and somewhat irregular at 48h.

Recognition of colonies in mixed cultures: N. gonorrhoeae (and *all* other neisseriae) give a positive oxidase test.

Method: pick suspect colonies onto filter paper moistened with 1 per cent solution of tetramethyl-*p*-phenylenediamine (the 'oxidase reagent'). Oxidase-positive colonies rapidly turn the paper a dark purple colour. Alternatively the plate may be flooded with oxidase reagent and observed for the development of dark-purple colonies: these should be sub-cultured for further tests without undue delay.

Identification: by carbohydrate utilization tests; *N. gonorrhoeae* produces acid from glucose only and no other sugar substrate.

Antigenic structure. Complex, cross-reactivity with other neisseriae: antigens in pili, cell wall lipopolysaccharide and outer membrane proteins; the latter may be strain-specific. Varies with cultural conditions.

Typing: although no definitive methods for subtyping are available, antibodies to gonococci raised in rabbits, and adsorbed to remove cross-reacting antibodies are of value in *identifying* a neisseria as *N. gonorrhoeae*. A diagnostic slide agglutination test with commercially-prepared antisera may be used in identification often in conjuction with carbohydrate utilisation tests.

Pathogenicity

The cause of the venereal disease gonorrhoea, a pyogenic infection of the urethra and, in females, of the uterine cervix.

Viability: dies rapidly outside the human host but may remain viable in pus for some time.

Antibiotic sensitivity

Sensitive to penicillin, ampicillin, tetracycline, macrolides, spectinomycin, cefuroxime and other drugs. Low-level penicillin resistance is now present in many strains isolated in Britain but this does not preclude treatment with the drug. In 1976, however, highly resistant ß-lactamase producing strains appeared abroad and are now increasingly and not uncommonly found in the United Kingdom.

NEISSERIA MENINGITIDIS

Habitat: the human nasopharynx: present in 5-10 per cent of normal people.

Laboratory characteristics

Morphology and staining: as for *N. gonorrhoeae.* Films of the CSF in meningococcal meningitis have a similar appearance to genital tract exudate in gonorrhoea but organisms are scantier.

Culture: requirements very similar to those described for *N. gonorrhoeae* although somewhat less exacting. On culture, the colonies are slightly larger. The use of an antibiotic-containing selective medium facilitates isolation of *N. meningitidis* from the normal mixed pharyngeal flora. The oxidase test is an aid to the recognition of suspect colonies but note that the commensal neisseriae will also give a positive reaction.

In meningitis the organism is present in CSF in pure culture and selective media are not required for isolation.

Identification: by carbohydrate utilization; *N. meningitidis* produces acid from maltose and glucose.

Antigenic structure. Although antigens are shared with other neisseriae recognition of polysaccharide antigens has allowed differentiation into serological groups. There are three main groups (A, B, C) and five subsidiary groups (X, Y, Z, 29E and W135): not all isolates are groupable.

Typing: serological typing into the above groups by slide agglutination with specific antisera.

Pathogenicity

Cause of meningococcal meningitis: in those susceptible, spread from the nasopharynx results in a septicaemia which is usually followed by rapid involvement of the meninges. However, chronic meningococcal septicaemia without meningitis is a recognised, although rare, clinical entity.

Viability: dies quickly at room temperature outside the human host.

Antibiotic sensitivity

Sulphonamide sensitivity was formerly the rule but resistance is now common: in Britain, at least 10 per cent of strains are fully resistant and a further 20 to 60 per cent partially resistant.

Sensitive to penicillin, ampicillin, chloramphenicol, tetracyclines, macrolides and other drugs.

COMMENSAL NEISSERIAE

Neisseria pharyngis group (includes *N. flava, N. perflava, N. sicca* and *N. subflava.*)

Habitat: regularly present in the mucous membranes of the mouth, nose and pharynx; less frequently in the genital tract.

These neisseriae can be differentiated from the two pathogenic species because of their ability to grow: (i) on ordinary agar not supplemented with blood or serum; (ii) at 22°C (room-temperature); (iii) on primary isolation in the absence of CO_2.

Colonies: often pigmented (yellow to green) and rough.

Biochemical activity: most species produce acid from a number of carbohydrates.

BRANHAMELLA

BRANHAMELLA CATARRHALIS

Until recently classified as *N. catarrhalis* and shares many characters with the commensal neisseriae.

Colonies: colourless but often opaque and sometimes rough.

Biochemical activity: fail to produce acid from carbohydrates.

Pathogenicity

This nasopharyngeal commensal may be a secondary invader in respiratory tract infections e.g. exacerbations of chronic bronchitis.

Antibiotic sensitivity: a proportion of strains are ß-lactamase producers and resistant to penicillin and ampicillin.

MORAXELLA

Gram-negative short bacilli arranged in pairs end-to-end; strictly aerobic; non-motile, oxidase-positive; sensitive to penicillin. *Moraxella lacunata* causes a purulent conjunctivitis.

17

Bacillus

Members of the genus *Bacillus* are aerobic, sporing, Gram-positive, chaining bacilli. *Bacillus species* are ubiquitous saprophytes but one, *Bacillus anthracis,* is an important pathogen responsible for anthrax in animals and man.

BACILLUS ANTHRACIS

Habitat: infected animals but spores are found in soil and pasture contaminated with vegetative cells from dead and dying animals.

Laboratory characteristics

Morphology: large (4-8 × 1.5 μm), non-motile, rectangular bacilli usually arranged in chains; spores—oval and central—and not formed in tissue but develop after the organism is shed or if it is grown on artificial media: capsulated in the animal body and on laboratory culture under certain conditions: the capsule consists of a polypeptide of D-glutamic acid.

Staining: Gram-positive; spores can be stained by modified Ziehl-Neelsen method.

McFadyean's reaction: used to demonstrate *B. anthracis* in the blood of animals in a heat-fixed film stained with polychrome methylene blue.

Observe for blue bacilli surrounded by purplish-red amorphous material due to disintegrated capsules and indicating a positive reaction: diagnostic of *B. anthracis.*

Culture: aerobe and facultative anaerobe; grows readily on ordinary media over a wide temperature range (optimum 35 °C): best temperature for sporulation is lower, 25-30 °C.

Colonies are large, dense, grey-white, matt and irregular: they are composed of parallel chains of cells and this gives the margin of the colony the so-called 'medusa head' or 'curled hair lock' appearance.

Blood agar: there is only slight haemolysis round the colony—a differential feature because other *Bacillus species* are markedly haemolytic.

Broth cultures develop a thick pellicle.

Gelatin stab cultures show growth along the track of the wire with lateral spikes longest near the surface—the 'inverted fir tree'; liquefaction is late, starting at the surface.

Antigenic structure: the antigenic components described include:
(i) a complex group of toxins
(ii) the capsular polypeptide

Pathogenicity

The cause of *anthrax.*

A wide range of animal hosts are susceptible; infection is characteristically septicaemic with splenic enlargement. Man is infected from animals or animal products.

Viability: vegetative cells are readily destroyed by heat but spores demonstrate a variable but often high level of heat resistance—in the dry state, up to 150°C for 1 h. Spores can remain viable for years in contaminated soil.

Antibiotic sensitivity

B. anthracis is susceptible to many antibiotics: penicillin is the drug of choice.

OTHER BACILLUS SPECIES ('ANTHRACOID BACILLI')

Habitat: saprophytes in soil, water, dust and air and on vegetation.

Many species are recognised: some (e.g. *B. megatherium, B. cereus*) are large-celled like *B. anthracis,* others (e.g. *B. subtilis*) are small-celled and shorter and thinner with rounded ends. Saprophytic species differ from *B. anthracis* in being motile, non-capsulated and in the distinct zone of haemolysis round colonies on blood agar: furthermore, they fail to produce a fatal septicaemia in laboratory animals.

The spores of certain anthracoid bacilli e.g. *B. stearothermophilus* or *B. megatherium* are used as a test of the efficiency of sterilisation by steam under pressure (the autoclave), by ethylene oxide and by ionizing radiation.

Bacillus cereus is a cause of food-poisoning (e.g. when contaminating rice).

18

Clostridium

Clostridia are anaerobic, sporing, Gram-positive bacilli. Most species are soil saprophytes but a few are pathogens. The most important are listed with some of their principal properties in Table 18.1.

Table 18.1 Pathogenic properties of the main medically-important species of clostridia

Species	Disease
Cl. perfringens	Gas gangrene, food poisoning
Cl. novyi	Gas gangrene
Cl. septicum	Gas gangrene
Cl. histolyticum	Secondary role in gas gangrene
Cl. fallax	Secondary role in gas gangrene
Cl. bifermentans	Secondary role in gas gangrene
Cl. difficile	Antibiotic-associated colitis
Cl. sporogenes	Doubtful pathogenicity in gas gangrene
*Cl. tetani**	Tetanus
Cl. botulinum	Botulism

*Forms round, terminal spores; the other species have oval, central, subterminal or terminal spores.

Habitat: human and animal intestine; soil, water, decaying animal and plant matter.

Laboratory characteristics

Morphology and staining: large (3-8 × 0.5 μm) rods, sometimes pleomorphic, filamentous forms common: Gram-positive but may stain irregularly or be Gram-negative in older cultures.

Spores: all species form endospores which are usually 'bulging', i.e. wider than the bacterial body; sometimes useful in identification e.g. *Cl. tetani.* Note that *Cl. perfringens* (the most common human pathogen) forms spores with difficulty.

Motile: with peritrichous flagella (*Cl. perfringens* is non-motile).

Capsule: Cl perfringens has a capsule but most are non-capsulated.

Culture:

1. Blood agar anaerobically: in mixed culture, addition of an aminogly-coside makes an excellent selective medium for clostridia.
2. Robertson's meat medium.

Anaerobic requirement: variable. *Cl. tetani* and *Cl. oedematiens* are exacting anaerobes; *Cl. perfringens* and *Cl. histolyticum* can grow in the presence of limited amounts of oxygen.

Colonial morphology: the main human pathogenic species show three types of colonial morphology.

a. *Cl. perfringens:* round opaque colonies: usually surrounded by a zone of β haemolysis on blood agar.
b. *Cl. tetani: Cl. septicum:* irregular, translucent colonies with thin spreading edges on the surface of moist agar: marked tendency to swarm especially with *Cl. tetani.*
c. *Cl. oedematiens: Cl. sporogenes: Cl. botulinum:* matt to glossy colonies the centre of which may be raised: margins irregular with filamentous rhizoid outgrowths: limited swarming may take place.

Biochemical activity

Saccharolytic: many species ferment sugars; this produces redden-ing of the meat particles in Robertson's meat medium with a rancid smell.

Proteolytic: production of enzymes that digest proteins is a common property of many clostridia: this causes blackening and digestion of the meat particles in Robertson's meat medium with a foul smell; although most clostridia liquefy gelatin, only those with considerable proteolytic activity are able to liquefy coagulated serum and egg.

Toxins: medically-important clostridial species produce several toxins: the exotoxins of *Cl. tetani* and *Cl. botulinum* are amongst the most toxic substances known: clostridial toxins are often lethal for various species of laboratory animal.

The range of saccharolytic and proteolytic activity in conjunction with tests for lecithinase and lipase activity and animal pathogenicity have been used in classification.

Identification: difficult: the characteristics and identification of four important pathogenic clostridia—namely *Cl. perfringens, Cl. tetani, Cl. botulinum* and *Cl. difficile*—are described below.

Antibiotic sensitivity

Sensitive to penicillin, metronidazole, and other antibiotics such as clindamycin, tetracycline, erythromycin: resistant to aminoglycosides.

CLOSTRIDIUM PERFRINGENS (CLOSTRIDIUM WELCHII)

Morphology: a short stubby bacillus in which spores are hardly ever seen.

Culture: most strains grow well on blood agar anaerobically producing ß-haemolytic colonies but some strains are non-haemolytic.

Biochemical activity: mainly saccharolytic: in tube cultures of litmus milk a characteristic 'stormy clot' is formed due to the production of acid and large amounts of gas.

Typing: Cl. perfringens can be divided into five types—A, B, C, D and E—on the basis of the twelve toxins formed; all five types produce α toxin. Type A is the human pathogen: the other types are important pathogens of domestic animals.

Toxins

Alpha (α) toxin: an enzyme, phospholipase C, which causes cell lysis due to lecithinase action on the lecithin in mammalian cell membranes.

Other toxins: include collagenase, proteinase, hyaluronidase, deoxyribonuclease. Several have haemolytic activity: some are described as 'necrostising' or 'lethal' from their effects on laboratory animals.

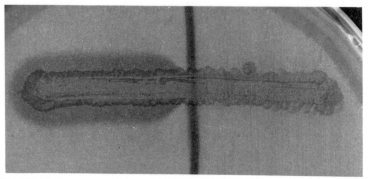

Fig. 18.1 Nagler reaction. The alpha toxin (a lecithinase) of *Cl. perfringens* has produced opacity due to degradation of lecithin in the medium on the left. This action has been neutralised by the antitoxin on the right of the plate.

Identification

Nagler reaction: identifies *Cl. perfringens* by neutralisation of α toxin by specific antitoxin: colonies are streaked on agar plates containing egg yolk (egg yolk contains lecithin) half of the plate having been spread with antitoxin: a dense opacity is produced by the growth of *Cl. perfringens* on the untreated half of the plate but there is no opacity on the area with antitoxin (Fig. 18.1).

Pathogenicity

Food-poisoning: when ingested in large numbers some strains of *Cl. perfringens* produce an enterotoxin in the gut causing diarrhoea and other symptoms of food poisoning.

Gas gangrene: wounds associated with necrosis of muscle may become infected with *Cl. perfringens* and other clostridia causing a severe and life-threatening spreading infection of the muscles.

CLOSTRIDIUM TETANI

Morphology: a longer, thinner bacillus with round terminal spores giving characteristic 'drum-stick' appearance (Plate 6).

Toxin: a protein and exceedingly potent; two components.
1. tetanospasmin: neurotoxic
2. tetanolysin: haemolytic

Identification: toxin neutralisation tests:
1. *in vitro:* culture the test organism on a blood agar plate half spread with tetanus antitoxin. Observation of haemolysis inhibited by the antiserum makes a presumptive identification: confirm, if necessary, by mouse inoculation.
2. *in vivo:* a culture filtrate of the test organism is injected into mice, some of which have been protected by previous inoculation of tetanus antitoxin. In a positive result the unprotected animals die with typical tetanic spasms; protected animals survive.

Typing: there are 10 serological types of *Cl. tetani:* types I and III are commonest in the UK: all produce the same toxin.

Pathogenicity

The cause of *tetanus,* a classical toxin-mediated disease in which *Cl. tetani* in a wound elaborates the powerful neurotoxin which spreads and acts on the central nervous system causing severe muscle spasms.

CLOSTRIDIUM BOTULINUM

Toxin: protein and even more potent than that of *Cl. tetani,* the toxin of *Cl. botulinum* in the most active known poison; it acts by preventing release of acetylcholine at motor nerve endings in the parasympathetic system; destroyed in 2min. at 60°–90°C dependent on type.

Typing: Seven serotypes—A, B, C, D, E, F and G each with a serologically distinct, but pharmacologically similar, toxin.

Human botulism is usually due to types A, B and E.

Identification: by testing in mice for neutralisation of toxin in a culture, patient's serum or food sample by specific antitoxin. Some of the mice are injected with antitoxin to A, B and E toxins; mice inoculated with the appropriate antitoxin survive, the others become paralysed and die.

Pathogenicity: produces a rare form of 'food-poisoning' known as botulism in which the symptoms are neurological rather than intestinal; due to ingestion of pre-formed toxin in food contaminated with the organism.

CLOSTRIDIUM DIFFICILE

Habitat: apparently rare in faeces of healthy adults although regularly present in faeces of healthy infants: found with its toxin in the faeces of patients suffering from acute pseudomembranous colitis.

Identification:

1. *Isolation:* from faeces using selective media, e.g. cefoxitin-cycloserine-fructose agar: anaerobic incubation: cultures produce a characteristic 'dung-like' aroma and the irregular rough colonies fluoresce under ultraviolet light. Gas liquid chromatography of pure broth subcultures demonstrates a characteristic pattern of volatile fatty acids.

2. *Demonstration of toxin* by inoculation of cell cultures (e.g. Vero or human embryo lung cells) with broth culture of *Cl. difficile:* cell cultures containing antitoxin to the toxin of *Cl. sordellii*—a serologically similar species—are included.

Observe: for cytotoxicity which is neutralised in the cultures containing *Cl. sordellii* antitoxin. Not all isolates of *Cl. difficile* are toxigenic.

Note: this test can be carried out using a faecal filtrate to demonstrate directly the presence of toxin in faeces.

Pathogenicity

The cause of antibiotic-associated pseudomembranous colitis: also responsible for less severe diarrhoeal disease.

19

Bacteroides and other non-sporing anaerobes

The genera of non-sporing anaerobes are listed in Table 19.1.

Table 19.1 Non-sporing anaerobes

	Gram-positive	Gram-negative
Bacilli	Bifidobacterium	Bacteroides
	Propionibacterium*	Fusobacterium
	Eubacterium	Leptotrichia
Cocci	Peptococcus	Veillonella
	Peptostreptococcus	

*Described in Chapter 13.

ANAEROBIC GRAM-NEGATIVE BACILLI

The classification of these organisms is complex: Tables 19.2 and 19.3 show simplified lists of some of the main species.

Table 19.2 Bacteroides species

	Bacteroides	
	Group	
Fragilis	Melaninogenicus/ oralis	Asaccharolytic
B. fragilis	*B. melaninogenicus*	*B. asaccharolyticus*
B. vulgatus	*B. oralis*	*B. gingivalis*
B. distasonis		*B. ureolyticus*
B. ovatus		(*corrodens*)
B. thetaiotaomicron		

Table 19.3 Fusobacterium and leptotrichia species

Fusobacterium	Leptotrichia
F. necrophorum	*L. buccalis*
F. nucleatum	
F. necrogenes	

Habitat:

Colon: bacteroides are present in enormous numbers in the faeces (10^{10} per gram or more). Almost all (about 80 per cent) belong to the fragilis group but only a minority, 1 in 10, are *Bacteroides fragilis:* another 10 per cent belong to the asaccharolytic group.

Female genital tract: bacteroides are common in the cervix and vaginal fornices: about 80 per cent belong to the melaninogenicus/oralis group; 15 per cent belong to the asaccharolytic group; *B. fragilis* is uncommon in this site.

Mouth: always present in large numbers in the normal mouth. About 70 per cent of strains belong to the melaninogenicus/oralis group, the commonest species being *B. oralis:* another 20 per cent of strains are fusobacteria.

Laboratory characteristics

Morphology and staining: Gram-negative, non-motile, non-sporing bacilli: bacteroides are usually small, ovoid or short bacilli. Fusobacteria and *L. buccalis* tend to be long and spindle-shaped: pleomorphism is common. *L. buccalis* contains granules sometimes Gram-positive in young cultures.

Culture: strict anaerobes: require media enriched with blood or haemin: the growth of many strains is improved by the addition of menadione (vitamin K3).

1. *Fluid media:* Robertson's cooked meat broth, preferably enriched, e.g. with yeast extract: the medium should be boiled and promptly cooled before use to remove dissolved oxygen and so improve anaerobiasis.
2. *Selective media:* incorporation of antibiotics to which bacteroides are resistant aids isolation from mixed cultures, e.g. blood agar containing an aminoglycoside (neomycin, kanamycin, gentamicin) or an aminoglycoside and vancomycin.

Incubate: anaerobically with 10 per cent carbon dioxide (which enhances growth) for a minimum of 48 h or longer.

Observe:

1. *Fragilis group:* light grey, opaque or translucent colonies usually 1-2 mm in diameter after 48 h incubation.
2. *Melaninogenicus-oralis group:* slower growing than the fragilis group; *B. melaninogenicus* characteristically produces black or

brown pigmented colonies with haemolysis on blood agar; other species in this group are non-pigmented.

3. *Asaccharolytic group:* slower growing than the fragilis group; *B. asaccharolyticus* produces black or brown colonies on blood agar. *B. ureolyticus* produces pitting or corroding of the agar round its colonies.

4. *Fusobacteria:* some species produce dull granular colonies which may be rhizoid or irregular.

5. *L. buccalis* colonies are lobate or convoluted.

Identification: most anaerobes can be identified to generic level by examination of a Gram-stained smear and by gas chromatographical analysis of fatty acid end-products of glucose metabolism. *Bacteroides* and *fusobacterium species* can be identified by a series of tests including colonial morphology, biochemical tests, growth inhibition by bile salts and antibiotic resistance. Species differentiation is time-consuming and costly: most laboratories simply report isolation of bacteroides.

Pathogenicity

The most common isolate from clinical specimens is *B. fragilis* and thus this species seems to have a special pathogenic potential.

Bacteroides are important in abdominal and gynaecological (including puerperal) sepsis: they are usually found along with other organisms, notably coliform organisms. Fusobacteria and *L. buccalis* play an important role in dental sepsis and Vincent's angina in association with melaninogenicus/oralis strains and borrelia (Plate 12).

It appears that these combinations of bacteria potentiate the ability of each other to cause infection—*pathogenic synergy.*

Antibiotic sensitivity

Like other anaerobes these organisms are sensitive to metronidazole: many are also sensitive to the lincomycins, chloramphenicol and cefoxitin. Members of the fragilis group of bacteroides are penicillin-resistant, but many strains of the other groups are penicillin-sensitive. There is uniform resistance to the aminoglycosides.

BIFIDOBACTERIUM

Bifidobacterium is a genus of diverse, *non-pathogenic* bacteria. *Bifidobacterium bifidum* is the type species. Previously they were classified as 'anaerobic lactobacilli'.

Habitat: dominant members of the colonic flora of infants (classically the breast-fed) and common also in the adult gut; normally present in the human vagina.

Laboratory characteristics

Morphology and staining: pleomorphic Gram-positive bacilli characterised by club-shaped rods and by branching forms which are often Y-shaped; non-sporing, non-motile.
Culture: strict anaerobes.

ANAEROBIC COCCI

The classification of this heterogeneous group—which includes both Gram-positive and Gram-negative cocci—is confused. They must be clearly separated from microaerophilic and carbon dioxide-requiring cocci which in the past have often been incorrectly considered anaerobic. One simple differential test is sensitivity to metronidazole: truly obligate anaerobic cocci are metronidazole-sensitive, the others are resistant.

Several genera have been described: *Veillonella* (minute Gram-negative cocci), *Peptococcus* (clustering Gram-positive cocci), *Peptostreptococcus* (chaining Gram-positive cocci), but their validity has been questioned because of variability in the results of Gram staining and the morphological appearances both microscopic and colonial.

Habitat: commensals in the oropharynx, colon and female genital tract.

Pathogenicity

Local sepsis in mixed infections with other anaerobes (bacteroides and fusobacteria) and aerobes.

Lactobacillus

Members of this genus are widely distributed as saprophytes in vegetable and animal material (e.g. milk, cheese): others are common human and animal commensal parasites. Classification is complex and in dispute: characteristically species attack carbohydrates to form abundant acid and are tolerant of an acid environment (pH 3.0-4.0). The best recognised species is *Lactobacillus acidophilus.* Strictly anaerobic lactobacilli are now placed in a separate genus, bifidobacterium.

LACTOBACILLUS ACIDOPHILUS

Habitat: present in the mouth, gastrointestinal tract and female genital tract; most vaginal lactobacilli (Döderlein's bacilli) seem to be *Lacto. acidophilus.*

Laboratory characteristics

Morphology and staining; large, often thick (1-5 × 1 μm) Gram-positive bacilli with a tendency to form pairs in line, short chains and filaments, non-branching, non-motile, non-sporing.

Culture: grow best under microaerophilic conditions in the presence of 5 per cent CO_2 and at pH 6.0. Small colonies form on ordinary agar after 48 h incubation: growth is better in media enriched with glucose or blood.

Selective media: acid media, e.g. tomato juice agar (pH 5.0), support the growth of lactobacilli but inhibit many other bacteria.

Pathogenicity

Lactobacilli are associated with dental caries.

Legionella and related genera

Legionella is a recently discovered genus of unusual, even unique properties: the base composition of its DNA is distinct from that of all other bacteria. There are several species but *Legionella pneumophila* is the main, common human pathogen. All cause similar respiratory disease. (Table 21.1).

Table 21.1 Species of Legionella and related bacteria

Species	No. of Serogroups
Legionella	
L. pneumophila	10
L. micdadei	1
L. bozemani	2
L. dumoffi	1
L. feeleii	2
L. gormani	1
L. hackeliae	1
L. longbeachii	2
L. wadsworthii	1

LEGIONELLA PNEUMOPHILA

Habitat: an environmental organism found in soil and water (including domestic water supplies and air conditioning units).

Laboratory characteristics

Morphology and staining: slender rods or cocco-bacilli: Gram-negative (Plate 11) *but legionellae sometimes do not stain well.*

Culture:
1. *On artificial media:* a fastidious bacterium difficult to grow on

ordinary laboratory media: use either blood agar or charcoal yeast extract agar supplemented with L-cysteine and ferric pyrophosphate or α-ketoglutarate respectively.

Incubate for 21 days at 35-37°C in 5 per cent carbon dioxide; colonies usually appear in 3-5 days.

Examine for growth and, if suspect colonies are Gram-negative bacilli, identify by: production of a fluorescent brown pigment on a L-tyrosine medium, biochemical reactions, direct or indirect immunofluorescence.

2. *In the fertile hen's egg:* a sensitive culture method not often used now.
3. *Animal inoculation*

Inject: guinea pigs intraperitoneally.

Observe: signs of illness—fever, ruffled fur, watery eyes—over 3-5 days.

Examine: the peritoneal exudate, spleen and liver by immunofluorescence.

Isolation from contaminated material: concentration before culture can be achieved by treatment with hydrochloric acid, pH 2, to which *Legionellae* are resistant. Use antibiotic-containing fluid-enrichment and solid-selective media.

Note. Diagnosis is most often serological: however, better culture media are now increasing the rate of isolation from clinical material.

Antigenic structure: ten serogroups have been recognised.

Pathogenicity

The cause of *Legionnaires' disease*—a severe form of pneumonia—and less serious respiratory disease, e.g. Pontiac fever.

Antibiotic sensitivity

Sensitive to erythromycin, rifampicin.

22

Spirochaetes

Spirochaetes are helical organisms which share many properties with Gram-negative bacteria.

Habitat: most are free-living and non-pathogenic but a few are causes of important human disease.

Laboratory characteristics

Morphology and staining: spirochaetes have a unique helical structure: a central protoplasmic cylinder is bounded by the cytoplasmic membrane and a cell wall of similar structure to that of Gram-negative bacteria (Fig. 2.3, Chapter 2). Between a thin peptidoglycan layer and the outer membrane run the *axial filaments* now regarded as internal flagella. These vary in number with the spirochaete, are fixed at the extremities of the organism and meet to overlap in the middle of the cell. They constrict and distort the bacterial cell body to give rise to the typical helical structure (Fig. 22.1).

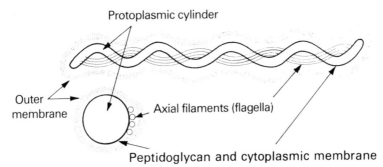

Protoplasmic cylinder

Outer membrane

Axial filaments (flagella)

Peptidoglycan and cytoplasmic membrane

Fig.22.1 The structure of a spirochaete.

The larger spirochaetes (e.g. *Borrelia species*) are Gram-negative: others stain poorly or not at all by the usual methods. They are too slender and too weakly refractile to be seen with the ordinary light microscope but can be rendered visible by dark-ground microscopy (Fig. 22.2) or by staining that enlarges their diameter, e.g. by deposition of heavy metals (Levaditi silver method) or by immunofluorescence.

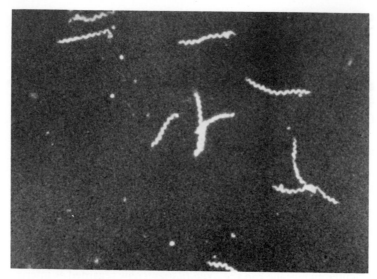

Fig.22.2 Dark ground photomicrograph of *Treponema pallidum.* (Reproduced with permission from Abbott Laboratories 'Slide Atlas of Infectious Diseases' 1982, Gower Medical Publishing, London. Photograph by Dr. R.D. Caterall.)

Motility is of three types:
1. Rotation about the long axis.
2. Flexion of the cells.
3. True movement, i.e. from one site to another.

Antibiotic sensitivity

Sensitive to penicillin and a number of other antibiotics; anaerobic spirochaetes are also sensitive to metronidazole.

GENERA

Five are recognised:
1. Treponema ⎤ Some species within these genera
2. Borrelia   ⎬ are associated with or pathogenic
3. Leptospira ⎦ for man and animals.

4. Spirochaeta—free living in the environment.
5. Cristispira—commensal, mostly in molluscs.

TREPONEMA

TREPONEMA PALLIDUM

Habitat: the disease lesions of primary and secondary syphilis.
Morphology: slender filamentous helices, 6-14 μm × 0.2 μm with 6-12 evenly spaced coils.
Culture: cannot be cultivated *in vitro* but can be propagated by inoculation of rabbits at certain sites, e.g. anterior chamber of eye, testes: this enables suspensions of *Tr. pallidum* to be prepared. These are used as the antigen to detect specific antibody in patient's serum.
Identification: in material from clinical lesions is by dark-ground microscopy.
Antigenic structure: all treponemes possess a common 'group' antigen; in addition, *Tr. pallidum, Tr. pertenue* and *Tr. carateum* share other antigens that make them similar but different from other treponemes.

Pathogenicity

Cause of the venereal disease syphilis.

Viability: a strict parasite that dies rapidly outside the body; it is very sensitive to drying and to heat.

TREPONEMA PERTENUE

The cause of yaws, a chronic relapsing non-venereal treponematosis widespread in the tropics, which is characterised by ulcerative and granulomatous lesions in skin, mucous membranes and bone.

TREPONEMA CARATEUM

The cause of pinta, a non-venereal treponamatosis affecting dark-skinned people in Central and South America: the skin becomes hyperkeratotic and depigmented.
Tr. pertenue and *Tr. carateum* are morphologically indistinguishable from and antigenically very similar to *Tr. pallidum*. They cannot be cultivated *in vitro*.

OTHER TREPONEMES

A number of species are found as commensals in the mouth, genital secretions and intestine. Their presence may create diagnostic problems when examining dark-ground preparations for *Tr. pallidum*. Some can be grown *in vitro:* they are anaerobes.

BORRELIA

BORRELIA VINCENTI

Habitat: the oro-pharynx, as a commensal and potential pathogen.
Morphology and staining: large spirochaetes 5-15μm$\times$0.5 μm with 3–8 irregular open coils; Gram-negative.
Culture: can be grown, with difficulty, in serum-enriched media; a strict anaerobe.
Identification: in exudates from clinical lesions by the appearance of a Gram-stained film.

Pathogenicity

In association with anaerobic fusiform bacilli responsible for gingivo-stomatitis and Vincent's angina.

BORRELIA RECURRENTIS

The cause of epidemic louse-borne relapsing fever.

BORRELIA DUTTONI AND OTHER SPECIES

The causes of endemic tick-borne relapsing fever.

Both diseases are encountered in parts of Asia, Africa and South America.

The diseases are characterised by febrile episodes alternating with afebrile periods and last for several weeks. Each relapse is the result of a change in the antigenic structure of the organism: antibodies already formed are ineffective against the new variants.

LEPTOSPIRA

Two species are recognised: *Leptospira biflexa* subdivided into a number of serogroups and *Leptospira interrogans* classified in some 20 serogroups each with one or more serotypes.

Habitat: Leptospires are found in moist environments. *L. biflexa* is a saprophyte present in pools, ditches and streams. *L. interrogans* is a potential pathogen for man and animals harboured in the kidneys of some rodents and domestic animals.

Morphology: spiral organisms 5-20 μm $\times$0.1 μm with very numerous closely-set coils and hooked ends.

Culture: grow readily in enriched fluid or semi-solid media under aerobic conditions; optimum temperature is around 30°C.

Identification: species, serogroup and serotype is by antigenic analysis.

Pathogenicity

L. interrogans is the cause of the zoonotic disease leptospirosis.

Viability: pathogenic strains may survive for days outside the animal body in moist surroundings as long as they are not acid.

23

Mycoplasma

Mycoplasmas are bacteria which lack cell walls. They resemble L-forms of bacteria but unlike them are independent naturally-occurring microorganisms.

Species: Table 23.1 lists some of the better-known mycoplasmas and their habitat.

Laboratory characteristics

Morphology and staining: pleomorphic; several different forms exist varying from small spherical shapes—which can pass filters—to longer branching filaments; Gram-negative but stain poorly with Gram's stain; colonies on agar are best stained with Diene's stain.

Culture: on semi-solid enriched medium containing 20 per cent horse serum, yeast extract and DNA; incubate aerobically for 7-12 days with CO_2 or in nitrogen with added CO_2.

Observe for typical 'fried-egg' colonies embedded into the surface of the medium.

Note: T-strain mycoplasmas form minute colonies ('T'=tiny) and are now classified in the genus *Ureaplasma;* some mycoplasmas with less exacting growth requirements have been assigned to a separate genus, *Acholeplasma.*

Identify by inhibition of growth round discs impregnated with specific antisera.

Table 23.1 Mycoplasma species

Species	Habitat
M. pneumoniae	Human respiratory tract
M. orale	Human mouth
M. salivarium	
M. hominis	Human genital, and possibly repiratory tracts
Ureaplasma urealyticum	Human genito-urinary tract
Acholesplasma laidlawii	Soil, water

Antigenic structure: although there is some antigenic sharing in complement fixation tests, mycoplasma species are distinct in tests of growth inhibition. *Note:* mycoplasma are like viruses in that their replication is inhibited by specific antibody without complement; this is not the case with bacteria.

Pathogenicity

Mycoplasma pneumoniae

The only member of the group of which the pathogenicity is unequivocally established: it is a major respiratory pathogen responsible for the important disease:
Primary atypical or 'virus' pneumonia.
It also causes febrile bronchitis and milder upper respiratory infections.

Ureaplasma urealyticum (T-strain mycoplasma)

Implicated in although not the major cause of non-specific urethritis or genital infection.

Mycoplasma hominis

The other genital species; has been implicated in some cases of gynaecological or past-partum sepsis.

Antibiotic sensitivity

Sensitive to tetracycline—the drug of choice for treatment: also, especially in children, erythromycin. Not surprisingly, mycoplasmas are resistant to antibiotics that interfere with bacterial cell wall synthesis, e.g. penicillin.

Tissue culture

Mycoplasmas are a common contaminant of cell lines, their persistence favoured by the presence of penicillin and other antibiotics in the tissue culture media. The most frequently isolated species are *M. hominis* and *M. orale*—probably derived from the mouth of those handling the cells. *Acholeplasma laidlawii* is also occasionally isolated.

Animal mycoplasmas

Mycoplasmas are also common in domestic animals e.g. cattle, goats, pigs, sheep and chickens.

Bacterial disease

24

Normal flora

The normal human body has a profuse bacterial flora: this consists of *commensal* bacteria which, although parasitic, exist in a symbiotic equilibrium with the host. Many of the commensals are *potential pathogens* and if the balance is disturbed by some breach of the body defences or if a parasite of increased pathogenicity is acquired, infection may result. Nevertheless, virtually all 'unequivocal pathogens' may be encountered in healthy carriers.

Change and composition. The normal flora is not static: although a basic flora persists, it is subject to constant change and individual components wax and wane; the reasons for the abundance of certain bacteria at particular sites are not understood.

Beneficial role. The presence of the normal flora prevents other more pathogenic bacteria from gaining a foothold in the body. The gut bacteria seem to be responsible for the normal structure and function of the intestine: they degrade mucins, epithelial cells and carbohydrate fibre and their metabolism produces vitamins, especially vitamin K.

Alteration by antibiotics. Broad-spectrum drugs disrupt the composition of the normal flora by inhibiting sensitive organisms and allowing overgrowth of resistant bacteria. As a rule, the host can cope with these changes but they occasionally result in serious infection.

Bacteriologists require a detailed knowledge of the normal flora: many specimens cultured in the laboratory yield commensal bacteria and they require to be distinguished from the pathogens responsible for the infection. Interpretation of the results of culture requires both knowledge and experience.

Distribution. With the exception of the alimentary tract, the internal organs of other systems are sterile in health, e.g. the bladder and kidneys, the bronchi and lungs, the CNS. Effective local defence mechanisms exist to maintain the sterility of these sites: in addition, chemical substances in serum and tissue fluids, e.g. complement, antibody, promote the powerful phagocytic activity of the polymorphonuclear leucocytes.

Below are the different sites of the body with the main bacterial species which make up their normal flora.

RESPIRATORY TRACT

The lower respiratory tract is sterile but the upper tract is colonised — heavily in the case of the mouth and nasopharynx (Table 24.1). Saliva contains about 10^8 bacteria per ml: gingival-margin debris and dental plaque consist almost entirely of microorganisms.

Table 24.1   Main bacteria of the normal respiratory flora

Nose	Staphylococcus epidermidis	
	Staphylococcus aureus	
	Corynebacteria	
Oro-pharynx	Viridans streptococci	
	Commensal neisseriae	
	Corynebacteria	
	Bacteroides	Mainly B. melaninogenicus, B. oralis
	Fusobacteria ⎤ Spirochaetes ⎦	Especially around the teeth
	Lactobacilli	
	Veillonella and other anaerobic cocci	
	Actinomyces	
	Haemophilus influenzae ⎤ Streptococcus pneumoniae ⎦	The important potential pathogens
	Less common:	
	Streptococcus pyogenes	
	Neisseria meningitidis	

GASTROINTESTINAL TRACT

The oesophagus has a flora similar to that of the pharynx. The empty stomach is sterile due to gastric acid.

The normal flora of the duodenum, jejunum and upper ileum is scanty but the large intestine is very heavily colonised with bacteria (Table 24.2).

Faeces: Contain enormous numbers of bacteria which constitute up to one third of the faecal weight: the majority of these bacteria seem to be dead. The number of living bacteria is about 10^{10} per gram and almost all (99.9 per cent) are anaerobes: the anaerobic environment of the colon is maintained by the aerobic bacteria utilising any free oxygen. Bifidobacteria are Gram-positive bacilli similar to lactobacilli

Table 24.2   Bacteria of the large intestine

Bacteroides (mainly members of the fragilis group which outnumber *B. fragilis* itself)
Bifidobacteria
Anaerobic cocci
Escherichia coli
Streptococcus faecalis
Clostridia
Lactobacilli
Less common inhabitants:
  Klebsiella species
  Proteus species
  Enterobacter species
  Pseudomonas aeruginosa

but strict anaerobes: like lactobacilli they are virtually non-pathogenic: they and *Bacteroides species* are the dominant anaerobes. *Bacteroides fragilis* is a considerably rarer gut inhabitant than other species classified in the fragilis group but has much greater potential for pathogenicity.

Table 24.3   Main bacteria of the male and female genital tracts

Female	
Vulva	*Staphylcococcus epidermidis*
	Corynebacteria
	Escherichia coli and other coliforms
	Streptococcus faecalis
	Yeasts
Vagina	Lactobacilli (known as Doderlein's bacilli)
	Bacteroides (especially *B. melaninogenicus*)
	Streptococcus faecalis
	Corynebacteria
	Yeasts
Male and female distal urethra	*Staphylococcus epidermidis*
	Corynebacteria

GENITAL TRACT

For anatomical reasons the female genital tract is much more heavily colonised than that of the male. Normal vaginal secretions contain up to 10^8 bacteria per ml. The genital flora is shown in Table 24.3.

Note. The secretions of both male and female genitalia may contain *Mycobacterium smegmatis*—acid-fast bacilli which, if they contaminate urine specimens, can easily be mistaken for tubercle bacilli.

Mycoplasma. Strains of mycoplasma called T-strains or ureaplasma which form minute colonies on culture are commonly present as part of the normal genital flora of both sexes.

SKIN

Although the hard dry surface of the skin may seem at first sight to be less hospitable for bacteria than moist mucous membranes, the skin has a rich resident bacterial flora (estimated at 10^4 organisms per cm^2). It is not evenly distributed: the bacteria exist in microcolonies of 10^2-10^3 organisms.

Anaerobic organisms predominate—particularly in areas with many serbaceous glands where anaerobic conditions prevail. In moist skin e.g. the axilla and groin, coliform organisms are often present (Table 24.4).

Table 24.4　Main bacteria of the skin flora

Propionibacterium acnes Anaerobic cocci	
Staphylococcus epidermidis Micrococci Corynebacteria	
Less common:	
Staphylococcus aureus	This potential pathogen is present in about 50 per cent of normal adults
Coliforms	

External auditory meatus

An extension of the skin and often profusely colonised: the main species found are *Staphylococcus epidermidis* and corynebacteria; acid-fast mycobacteria are occasionally present in the wax.

Conjunctival sac

Bacteria are scanty: occasional *Corynebacterium xerosis* and *Staph. epidermidis.*

25

Host-parasite relationship

Infection is the result of breakdown in the host-parasite relationship and follows when the balance is tipped in favour of the parasite.

Man 'the host' lives in general balance with his environment. The environment includes numerous bacteria found in all sites, animate and inanimate, with which man comes in contact: most important is his own normal flora.

Bacterial disease is mostly due to organisms which form—at least from time to time—part of the commensal flora: naturally there are exceptions and some pathogenic bacteria are never commensal and are found only in disease.

Selection pressure favours the survival of bacteria with limited pathogenicity which can maintain a symbiotic relationship with their host. Virulent pathogens which severely incapacitate or kill man are denied the opportunity to spread within a community because their host is no longer able to circulate and come in contact with other individuals: human beings are gregarious but only when they are healthy.

DEFENCE MECHANISMS OF THE HOST

Host factors influence the outcome of host-parasite interaction. Some are linked with socio-economic status, e.g. malnutrition, poverty and overcrowding, conditions that also favour the transmission of a virulent pathogen within a community. Some other host factors are listed below.

1. *Nutrition:* malnutrition (e.g. vitamin or protein deficiency) predisposes to infection.
2. *Age:* the very young (especially preterm neonates) and the aged are particularly liable to infection.
3. *Sex:* rarely important: occasionally attributable to occupational risks.

4. *Race:* sometimes a factor e.g. blacks are more susceptible than whites to tuberculosis.
5. *Occupation:* some occupations (e.g. those associated with inhalation of minerals) have a higher than normal risk of infection with certain microorganisms (such as *Myco. tuberculosis*).
6. *Impairment of the host immune response by:*
   a. Treatment e.g. immunosuppressive, cytotoxic or steroid drugs, radiotherapy
   b. Disease e.g. malignancy (especially of the lymphoid system), metabolic diseases (diabetes, renal or hepatic failure)

The host has a number of defence mechanisms with which to counteract bacterial aggression
There are two categories of defence mechanisms:
1. *Non-specific:* not directed at a particular organism and non-immunological.
2. *Specific:* directed against a particular organism and these are dependent on immunological mechanisms.

NON-SPECIFIC DEFENCE MECHANISMS

1. *Skin*
Skin is a tough layer or integument which forms an excellent and generally impermeable barrier to invasion of the tissues by organisms either from the normal flora of the skin or the environment: infection is frequent when this barrier is breached, e.g. by a surgical or traumatic wound.
2. *Normal flora*
Can make it difficult for exogenous pathogens to establish themselves. Substances such as fatty acids with antibacterial activity are produced by skin flora from glycerides in sebum and by intestinal anaerobes from the contents of the colon.
3. *Lysozyme*
Lysozyme is an enzyme found in tears and other body fluids which lyses the mucopeptide of the cell wall of Gram-positive bacteria.
4. *Flushing action of:*
   Tears: with lysozyme, tears keep the surface of the eye sterile.
   Respiratory tract mucus: traps bacteria and constantly moves them upwards propelled by *cilia* on the cells of the epithelium.
   Urine: voiding helps to flush out bacteria that have gained entry to the bladder.

5. *Low pH in:*

Stomach: ingested bacteria are usually destroyed by the low pH of stomach acid; this can, of course, be buffered by food.

Vaginal secretions in young women have acid pH due to lactobacilli which metabolise glycogen present because of circulating oestrogens in the epithelium: the lactic acid produced prevents access of harmful bacteria.

Phagocytosis

Phagocytosis is the most powerful and most important of the non-immune defence mechanisms: it is mediated by scavenger cells which ingest invading organisms and destroy them intracellularly by enzymes.

There are two types of phagocyte:

1. *Neutrophil polymorphonuclear leucocytes* (the 'polymorphs')

Also known as *microphages.* They are produced in the bone marrow and, when mature, circulate in the bloodstream for 6-7 h. Short-lived cells arrive rapidly at the scene of infection attracted by chemotactic substances elaborated during the inflammatory process. Phagocytosis is promoted by specific antibody and complement which act as *opsonins.* Polymorphs act as an early defence against infection and are the 'pus cells' seen in the exudate from acute infections.

2. *Macrophages of the mononuclear phagocytic* (or *reticulo-endothelial*) *system*

Produced in bone marrow, they travel as *monocytes* in the blood stream to become distributed as *free macrophages* (in lung alveoli, the peritoneum and inflammatory granulomas) or *fixed macrophages* integrated into the tissues (in lymph nodes, spleen, liver (Kupffer cells), CNS (microglia) and connective tissue (histiocytes)). Phagocytosis by these long-lived cells can either be non-specific or promoted by antibody and complement.

Macrophages process bacterial antigens and present them to lymphocytes to stimulate a specific immune response: they also play an important part in cell-mediated immunity.

Phagocytic function can be divided into four stages:

1. *Chemotaxis:* attraction of the phagocyte to the site of the organism.

2. *Attachment (adherence)* of the bacterium to the membrane of the phagocyte.
3. *Ingestion* in which the phagocytic cell extends small pseudopods to envelop the bacterium: these fuse to form a pouch or *phagosome*. Lysosomes containing hydrolytic enzymes and other bactericidal substances migrate towards the phagosome and fuse with its membrane to form a *phagolysosome*.
4. *Intracellular killing of the ingested bacterium:* most bacteria are killed within a few minutes of phagocytosis although the degradation of the bacterial cell may take several hours.

Complement

A family of proteins present in serum: these react together one after another in a cascade once the reaction has been triggered. Activation of the first stage takes place when specific antibody combines with a bacterial or other antigen: the sequential reaction liberates fragments that attract phagocytic cells (chemotaxis), promotes subsequent phagocytosis and induces the changes characteristic of the inflammatory reaction.

SPECIFIC DEFENCE MECHANISMS—THE IMMUNE SYSTEM IN INFECTION

There are two main mechanisms by which the host mounts a specific immune response against bacterial infection.
1. *The humoral (antibody) response*
2. *The cell-mediated response*

Antibody response

Antibodies are proteins in the blood stream which are produced in response to infection by microorganisms: they are specifically directed against the bacterium or its component parts—usually protein or occasionally carbohydrate in nature. Most are *antigens* and stimulate antibody production.

B-(bone marrow derived) lymphocytes. When an antigen—e.g. on a bacterium—encounters B-lymphocytes in the spleen or lymph nodes, the lymphocytes are activated and changed into antibody-secreting plasma cells. The antigen is presented by macrophages and involvement of T-lymphocytes is required to initiate the immune response to some antigens.

Antibodies are *immunoglobulin* (Ig) high molecular weight protein molecules: their structure is Y-shaped and consists of a *Fc fragment* (the stem of the Y) and two *Fab fragments* (the arms of the Y): the Fab fragments contain the combining sites for specific antigens and, in antibodies to different antigens, show highly variable amino acid sequences. The Fc fragment of different antibodies, on the other hand, has a relatively constant amino acid composition and is the site for attachment of complement.

Although there are five types of immunoglobulin only three are concerned in the response to infection. These are described below:

1. *IgM:* a pentamer of IgG of 1×10^6 molecular weight—the first antibody produced: appears approximately a week after infection and persists for about 4 to 6 weeks.
2. *IgG:* 1.6×10^5 molecular weight monomer—the main antibody produced: appears about 2 weeks after infection but persists for long periods of time.
3. *IgA:* 1.7×10^5 molecular weight, a monomer in blood, present as a dimer in body secretions e.g. saliva, respiratory and alimentary mucus, tears, colostrum, etc. IgA in extracellular fluids (secretory IgA) is coupled to a carbohydrate transport piece which is not found on serum IgA.

Mechanism of action of antibodies

Antibodies are powerful defence mechanisms against viruses since they neutralise viral infectivity: they are much less effective on their own (i.e. without complement) against bacteria.

Nevertheless antibodies are important in combating bacterial infection by the following mechanisms

Neutralisation of toxins

Promotion of phagocytosis: antibody-coated bacteria are more readily phagocytosed, i.e. they are opsonised more effectively than by complement alone.

Bacterial lysis: certain Gram-negative bacilli are lysed in the presence of antibody and complement.

Cell-mediated immunity

Delayed hypersensitivity—or cell-mediated immunity—was first described in tuberculosis in the late nineteenth century: the mechanism, however, has only recently been discovered. Delayed hypersensitivity is especially important in infections due to organisms which persist or multiply intracellularly such as the bacteria which cause tuberculosis, leprosy, brucellosis and viruses. In delayed hypersensitivity, the inflammatory lesion is infiltrated with sensitised T-lymphocytes.

T-(thymus dependent) lymphocytes are a population of lymphocytes developed in the thymus. Responsible for cell-mediated immunity they comprise 75 per cent of the circulating lymphocytes in man: when sensitised or primed T-lymphocytes encounter the specific antigen with which they can react, they undergo transformation to become actively metabolising blast cells and release lymphokines.

Lymphokines have the following activities:

1. *Inhibition of macrophage migration* (this probably localises the macrophages to the site of infection)
2. *Chemotactic attraction* of lymphocytes, macrophages and polymorphonuclear leucocytes to the site of infection
3. *Increase in capillary permeability*
4. *Mitogenic activity:* or stimulation of normal, i.e. unsensitised lymphocytes to divide

The overall effect of delayed hypersensitivity is to limit the size of the lesion and surrounding inflammatory reaction and to localise the organism within it.

AGGRESSIVE MECHANISMS OF THE PARASITE

Bacteria vary in their *pathogenicity*—or ability to produce disease—in man.

Virulence is a commonly used but ill-understood term which indicates the degree of pathogenicity.

Neither pathogenicity nor virulence is easy to measure: it is impossible to do so in human beings for ethical reasons; experiments in laboratory animals can measure the incidence of disease or death following inoculation of an organism but are not always—and probably not often—analogous to the behaviour of the organism in the human host.

Bacteria as pathogens have two basic mechanisms of producing disease:

1. *Invasiveness*
2. *Toxin production*

Although bacteria can cause disease which is predominantly invasive or toxic in origin, most infections are due to a combination of both activities.

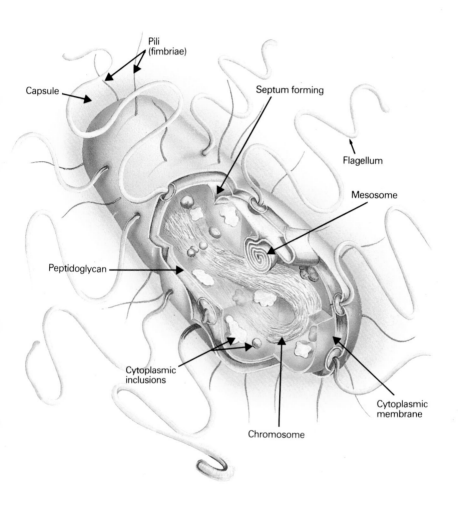

Pili
(fimbriae)

Capsule

Septum forming

Flagellum

Mesosome

Peptidoglycan

Cytoplasmic
inclusions

Cytoplasmic
membrane

Chromosome

Plate 1
Diagram of bacterial cell.

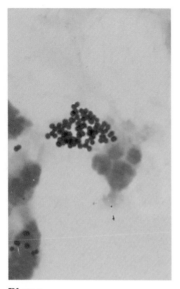

Plate 2
Staphylococcal pus (approx.
× 1000).

Plate 3
Streptococcal pus (approx. × 1000).

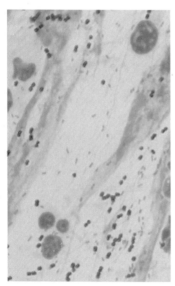

Plate 4
Sputum: pneumococcus and
haemophilus (approx. × 1000).

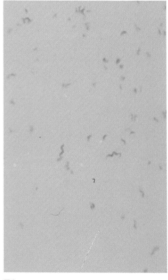

Plate 5
Campylobacter (approx. × 1000).

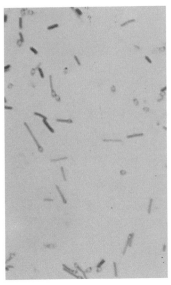

Plate 6
Clostridium tetani (approx. × 1000).

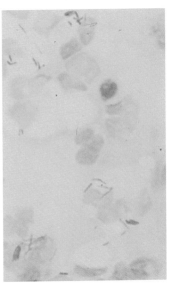

Plate 7
Mycobacterium tuberculosis:
film (approx. × 1000).

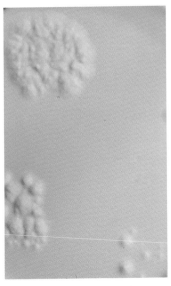

Plate 8
Mycobacterium tuberculosis: culture
(approx. × 1000).

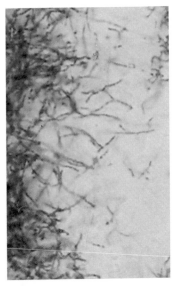

Plate 9
Sulphur granule of actinomycosis
(approx. × 1000).

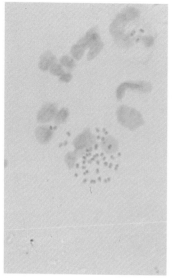

Plate 10
Gonococcal pus (approx. × 1000).

Plate 11
Legionella pneumophila (approx. × 1000).

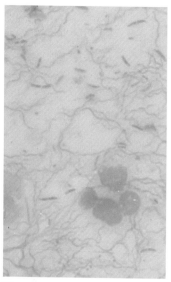

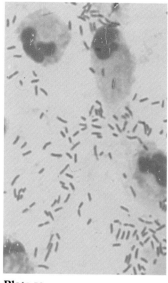

Plate 12
Vincent's angina: film (approx. × 1000).

Plate 13
Urinary deposit: coliform infection (approx. × 1000).

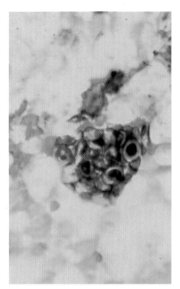

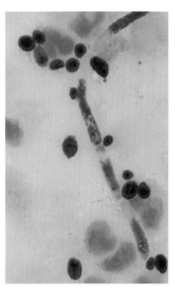

Plate 14
Pneumocystis carinii in bronchial washing (approx. × 1000).

Plate 15
Candida albicans in pus (approx. × 1000).

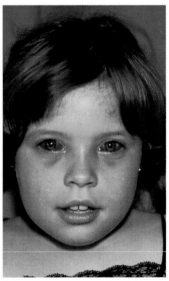

Plate 16
Pertussis: sub-conjunctival haemorrhages due to spasms of severe coughing (Photograph by Dr A K R Chaudhuri).

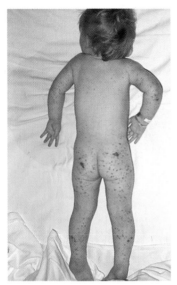

Plate 17
Meningococcal septicaemia:
haemorrhagic rash (Photograph by
Dr D H M Kennedy).

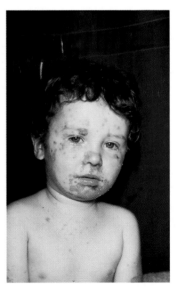

Plate 18
Impetigo (Photograph by Dr W C
Love).

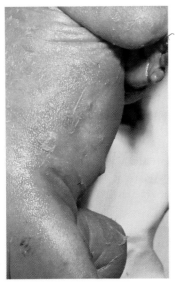

Plate 19
Ritter-Lyell disease: 'scalded skin'
produced by epidermolytic toxin of
Staphylococcus aureus (Photograph
by Dr W C Love).

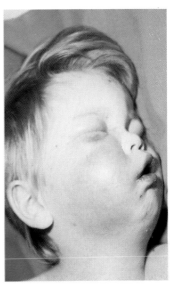

Plate 20
Acute streptococcal cellulitis
(Photograph by Dr A K R
Chaudhuri).

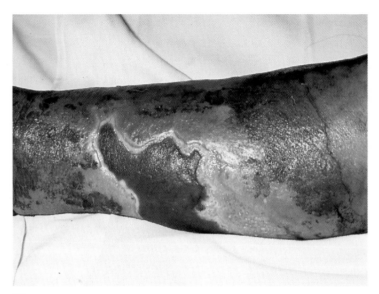

Plate 21
Progressive bacterial gangrene due to *Streptococcus pyogenes* and *Bacteroides species* (Photograph by Dr D H M Kennedy).

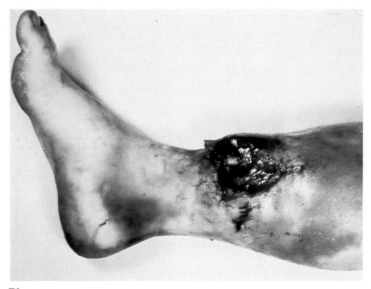

Plate 22
Clostridial gas gangrene complicating a compound fracture (Photograph by Professor J G Collee).

Invasiveness

Invasiveness is the ability of an organism to spread within the body once it has gained its initial foothold. It depends on (i) the action of *toxins* elaborated by the bacterium and (ii) *cell surface components* which enable it to resist phagocytosis. The latter may be demonstrable as visible capsules (e.g. the pneumococcus) or present as part of the cell wall (e.g. the M protein of *Strep. pyogenes*, the K antigens of enterobacteria).

A high degree of bacterial invasiveness is usually associated with severe infection: spread from a local site is often via the lymph channels *(lymphangitis)* to the draining lymph nodes *(lymphadenitis)* and possibly then to the blood stream *(septicaemia)*—one of the most serious manifestations of infection.

Bacterial toxins

These are of two types:
1. *Exotoxins*
Liberated extracellularly from the intact bacterial cell (and also produced on cell lysis) exotoxins spread via the blood stream or sometimes, nerves: they can produce ill effects locally and also at sites far distant from the infective process. A few bacterial diseases (e.g. diphtheria, tetanus) are the result of microorganisms which remain localised at the site of entry but form exotoxins which produce severe, distant effects. Other exotoxins are enterotoxins (e.g. produced by *Staph. aureus*) and botulinum, the neurotoxin produced by *Cl. botulinum*.

2. *Endotoxins*
Endotoxins are O antigens, structural components of the cell wall of Gram-negative bacteria and liberated only on cell lysis or death of the bacterium. Although differing antigenically they all produce the same physiological effects.

The main differences between exotoxins and endotoxins are shown in Table 25.1.

Some bacteria produce or contain substances often called 'aggressins' which enable them to withstand the host defences:

phospholipase— cytolytic
coagulase   — deposits fibrin
hyaluronidase— dissolves cell-binding material, aids spread

Table 25.1 Bacterial toxins

	Exotoxins	Endotoxins
Composition	Protein	Lipopolysaccharide
Action	Specific	Non-specific
Effect of heat	Labile	Stable
Antigenicity	Strong	Weak
Produced by	Gram-positive, some Gram-negative bacteria	Gram-negative bacteria
Convertibility to toxoid*	Yes	No

*Toxoid is toxin treated, usually with formaldehyde, so that it loses toxicity but retains antigenicity.

CONCLUSIONS

The host-parasite relationship is therefore complex and delicately balanced: the defence mechanisms of the host protect against invasion or uncontrolled replication of commensal bacteria but permit a symbiosis with the normal flora.

The parasite, for its part, depends for long-term survival in human populations on causing minimal or no damage to the host. When this balance is disturbed internally, or a result of invasion from outside, bacterial disease results.

26

Epidemiology

Epidemiology is the study of the spread of infection. From the point of view of the origin of the infection, bacterial disease can be considered in two categories:

1. *Endogenous:* when the organism is derived from the individual's own flora; but *note*—an epidemic organism may first be acquired as part of the normal flora.

2. *Exogenous:* when the organism is acquired from outside sources.

Epidemiology is chiefly concerned with exogenously-acquired infection and how infectious disease affects a community or the population at large.

RESERVOIRS AND SOURCES

In most instances, the reservoir and the source of infection are one and the same—but not always: sometimes the source has acquired the infecting organisms from the reservoir—the reservoir remaining distant from the eventual victim of the infection. Infection may be acquired from numerous different reservoirs and sources. Below are some of the most common:

Human beings

By far the most important source of infection. Human beings act as sources of infection in three main ways:

1. Active cases of disease

Patients suffering from an infectious disease often shed the causal organism in large numbers; on the other hand, they may be so incapacitated by illness that the illness prevents them from circulating in the community.

2. Inapparent (subclinical) infections

People with symptomless diseases continue to circulate in the community and, unless laboratory tests are carried out, are unrecognised as sources of infection.

3. Carriers

Some patients—some of whom may have had symptomless infection—become long-term carriers and excretors of pathogenic organisms; in some instances the organisms become part of the normal flora, in others they are shed from some chronic, perhaps symptomless, focus of infection; carriers also circulate in the general population and are often undiagnosed.

Animals

Infection is an occupational hazard to farm workers, veterinary surgeons and slaughtermen.

Animal products such as meat, milk, hides, etc. can be sources of infection to the general population also.

Food

An important source of infection; it may be infected at its animal origin (e.g. poultry, milk) or later become contaminated when handled by man. It is more than a passive vehicle; pathogenic bacteria multiply rapidly and in some instances produce toxins if the contaminated food is allowed to remain at the environmental temperature.

Water

Britain has a safe and pure water supply which is uncontaminated by sewage. If this sewerage system breaks down or in countries (which include most of those in tropical or sub-tropical areas) where the water supply is often polluted, water-borne outbreaks of infection can ensue—usually due to faecal bacteria.

Soil

Most organisms in the soil are free-living bacteria which are non-pathogenic or of very low pathogenicity; however, soil may become contaminated with pathogenic organisms derived from animal faeces or discharges.

Air

Air has a resident flora of bacteria of relatively low pathogenicity; however, air bacteria are also derived from human beings, who shed organisms both from skin as desquamated scales and from the respiratory tract; air also contains dust particles.

Dust

Dust is also contaminated with bacteria shed from human beings and from clothing, furnishings, bed linen and so on; bacteria in the air (particularly if contained in droplets) are deposited or fall into the dust and, conversely, dust particles are swept up into the atmosphere by air currents or movement of personnel.

Fomites

These are, strictly speaking, objects of a porous nature which absorb and can pass on contagion: in practice, any object which can be contaminated with bacteria is regarded as a fomite.

It can be seen that many of these reservoirs and sources are interrelated and interdependent.

ROUTES OF INFECTION

The route of infection depends largely on the reservoir or source. The main routes are listed below.

Inhalation

A common route of infection—particularly of the respiratory tract. Spread is by inhalation of droplets of respiratory secretions from someone suffering from active or symptomless infection.

Droplets produced by sneezing and coughing can be:
1. *Large*—which travel only a few feet and contaminate the environment to produce infected dust and fomites *or*
2. *Small*—droplet nuclei (5 μm in diameter or less) remain suspended in the air; these droplets may be inhaled to reach the lower respiratory tract directly.

Ingestion

A common mode of spread: organisms derived from faeces may be passed on by the *faecal-oral route* usually via contaminated fingers, towels, etc.: faecal organisms may also spread indirectly via contaminated drinking water or food.

Contact

Direct contact can spread infection via hands, kissing; indirect contact from fomites usually involves final transfer via the hands.

Sexual transmission

Many infectious diseases spread wholly or partially by sexual intercourse; not surprisingly, such diseases usually consist of lesions on the genitalia, and the organisms—which often do not survive well outside the body—are inoculated directly onto mucous membrane.

Inoculation

Infection may take place through broken skin either due to accidental trauma, animal bites or as a result of surgery—a common route of infection; infection can be introduced by other medical procedures such as catheterization, injection, blood transfusion, etc.

Vector-borne

Arthropod insects and parasites which bite and are blood-sucking such as mosquitoes, ticks, lice and fleas are a route of some infections; a common route in tropical countries but rare in temperate countries like Britain.

Transplacental

Some maternal infections can cross the placenta to infect the fetus: less common with bacteria than viruses.

EPIDEMIOLOGY

Control and prevention of infectious disease, depends on an understanding of reservoirs, sources and routes of transmission: sometimes even when these are known it is impossible to control infection, for example, surgical wound infection, food-poisoning, most respiratory infections.

The spread of an infectious disease in a community depends on the following factors.

Number of susceptible hosts

The chance of an outbreak of infection correlates directly with the number of susceptible people: immediately after an outbreak the general level and incidence of antibody is high, i.e. there is good *herd immunity;* with time, the level of antibody wanes and more children — who have never experienced the infection (and therefore have no antibody at all) — are born into the population. The herd immunity then becomes low with a relatively large number of susceptible hosts; under these circumstances the organism can infect again on a large scale.

Geometric mean titre of antibody to a particular organism within the population is an indication of herd immunity: this is calculated from the mean of the logarithms of the titres in the serum samples tested (this method of calculation removes the bias created by one exceptionally high (or exceptionally low) titre of antibody in the sample).

Pathogenicity

Some organisms inherently possess a high capacity to spread (infectiousness or communicability) or to cause disease (virulence). Some organisms — like *Salmonella typhi* — can infect in very small doses: others — like *S. typhimurium* — require large numbers of organisms to establish infection; *S. typhi* is therefore more virulent than *S. typhimurium.* Influenza virus is the classic example of a microorganism with an extremely high infectiousness.

Route of spread

This factor may determine an outbreak of infection: for example, a breakdown in the pure water supply gives faecal organisms from sewage the opportunity to infect large numbers of people via contaminated water. Blood transfusion is mainly responsible for hepatitis B and the increase in male homosexual promiscuity for the current outbreak of acquired immune deficiency syndrome (AIDS).

Carriers

When there is a relatively high proportion of carriers in a community this also increases the likelihood of an outbreak of infection. Outbreaks of meningococcal meningitis are often preceded by an increase in the proportion of people in the community who carry the organism in their throat.

Climate

Climate may contribute to the incidence of infection. Cold wet weather increases the number of exacerbations of chronic bronchitis whereas hot weather enhances the risk of food-poisoning.

MEASUREMENTS IN EPIDEMIOLOGY

Epidemiologists have various measurements by which infection in a community can be assessed: a community may be a family, an institution, a geographical area or the population of an entire country.

These measurements include:

Incubation period

To trace the spread of an outbreak it can be of great importance to know the incubation period—particularly if this is relatively long (e.g. about 2 weeks or longer): this may also allow time for preventive or containment measures to be instituted.

Incidence or incidence rate

The number of cases of disease in the community expressed as the ratio of the number of cases per 1000 people (or 10 000, 100 000 or per million—whatever is appropriate) in the population concerned; *prevalence* is a similar estimate but usually refers to the incidence in a population within a certain stated time.

Attack rate

Another way of expressing incidence: this term is used when the rate is applied to a particular defined group, for example, inhabitants of an institution.

Secondary attack rate

An important statistic for an epidemiologist, this is the number of secondary cases of infection which appear in the contacts—e.g. family or workmates, etc.—of an index case of the disease.

Mortality rate

The proportion of people in the community who die from the disease expressed as deaths per 1000, 100 000, etc.

Case fatality rate

The proportion of patients with the disease who die as a result of it: usually expressed as a percentage.

SURVEILLANCE

Measurements of infectious disease are constantly being monitored by appropriate community medical specialists. Outbreaks cannot always be prevented but preventive measures promptly applied can successfully halt the spread of an epidemic disease.

EPIDEMIC BACTERIAL DISEASES

Most worldwide *epidemics* or *pandemics* nowadays are viral but cholera is an exception and in the past 20 years cholera has again spread west from its original focus in the Far East: countries with adequate systems for sewage disposal are not at risk from this disease.

Many bacterial diseases are constantly present in the population and are a major problem for public health authorities. These include salmonella food-poisoning, gonorrhoea and dysentery. The reservoirs, sources and routes of spread of these diseases are well understood but it is impossible to control them.

Other bacterial diseases like streptococcal sore throat are always endemic in the population and wax and wane—usually for no apparent reason.

Immunisation can radically alter the epidemiology of an infectious disease—indeed, the main purpose of immunisation is to do this. Vaccines are generally less effective in the prophylaxis of bacterial diseases than of viral diseases. *Toxoids* are an exception to this and diphtheria and tetanus prophylaxis by toxoids is extremely effective.

Whooping cough vaccination is a controversial subject which illustrates a problem common to other bacterial diseases also. The vaccine has a protective effect but this is incomplete. Unfortunately it may also have serious side effects and can occasionally cause convulsions and brain damage. Although the incidence of these side effects is low, widespread publicity about some tragic cases resulted in a decline in the acceptance rate of triple vaccine: the widespread outbreak of whooping cough in 1978-79 was almost certainly at least partly a result of this.

Rare outbreaks such as that of typhoid fever in Aberdeen in 1964—which was due to a contaminated tin of corned beef—illustrate the need for constant vigilance by community medicine specialists.

27

Specimens for bacteriological investigation

Laboratory diagnosis in bacteriology depends on:
1. Careful collection of the appropriate *specimens*. These must be accurately *labelled* and if there is a *risk of serious infection* being transmitted by the specimen this must be indicated.
2. *Transport* of the specimen to the laboratory without delay.
3. *Laboratory investigation* to establish the cause of the infection.

SPECIMENS

Some examples are given below:
1. *Urine:* a mid-stream specimen taken with precautions to avoid contamination.
   Delay: if this is inevitable, store the sample at 4 °C or use a container with boric acid to prevent bacterial multiplication.
   Dip slide: a slide coated with medium is dipped into the urine and replaced in a sterile container for transport to the laboratory; delay is unimportant since the organisms are already inoculated onto solid medium.
   Tuberculosis: if suspected, collect three entire first morning specimens.
2. *Faeces:* collect in a plastic container; if not available, a rectal swab can be taken.
3. *Sputum:* a morning specimen in a wide-mouthed container.
   Tuberculosis: if suspected, collect specimens on three consecutive mornings.
4. *Serous fluids:* (e.g. pleural, synovial, ascitic fluids); collect in a sterile container with citrate to prevent clotting.
5. *Cerebrospinal fluid:* collect by lumbar puncture into a sterile container.
6. *Blood culture:* blood, aseptically collected, is injected into each of two screw-capped bottles with a perforation in the cap to allow

injection through the rubber liner; one bottle contains aerobic, the other anaerobic, medium.

7. *Clotted blood* (for serological tests, antibiotic assays): send 5–10 ml in a sterile container.

Swabs are widely used to collect samples from infected sites. They consist of a shaft (wooden, plastic, rigid paper or metal) with a cotton-wool tip which is rubbed over or inserted into the lesions and replaced in a stoppered tube (plastic or glass) for transport to the laboratory. Essential for sampling some areas, e.g. throat, cervix, but where *pus* is available, it is always better to collect this for examination.

LABELLING

All specimens must be labelled with the patient's name and ward (or home address) and accompanied by a *request form* giving other details. These should include the nature of the specimen, clinical history, antibiotic therapy, date and time of collection etc., and are essential for the interpretation of results.

RISK OF SERIOUS INFECTION

Specimens which may present a hazard to laboratory staff (e.g. blood samples positive for hepatitis B virus, sputum from a known case of open pulmonary tuberculosis, etc.) must be labelled **'Dangerous specimen'**.

Doctors must remember that they too may infect themselves when taking a specimen unless this is done carefully.

TRANSPORT

Most specimens need to be sent to the laboratory without delay: some bacteria die off quickly outside the body, others may overgrow and give a false impression of their original numbers.

If delay cannot be avoided specimens should be kept cool or at room temperature, except blood cultures and CSF which should be incubated at 37 °C.

Stuart's transport medium: a sloppy agar containing salts with a reducing agent and sometimes pieces of charcoal. Used with swabs: when swabs are placed in tubes containing this medium, delicate organisms are preserved.

LABORATORY INVESTIGATION

The methods used in the bacteriology laboratory are described in Chapter 5. Diagnosis of an infection can be achieved by:

1. *Direct demonstration:* presumptive diagnosis of certain infections can be made by detecting the causal organism morphologically in a smear from the specimen. Rarely possible with material collected on a swab—a specimen of pus, exudate or other fluid (e.g. CSF) should be sent to the laboratory to do this. Films stained by Gram's method can demonstrate a wide range of pyogenic bacteria and if acid- and alcohol-fast bacilli are seen in a Ziehl-Neelsen preparation an almost certain diagnosis of tuberculosis can be made. A positive finding is of great value because treatment (based on the likely sensitivity of the bacterium to antibiotics) can be started without delay. Many infections cannot be diagnosed in this way; for example, a Gram-stained film of faeces reveals, in almost all cases, a non-specific mixed flora and is of no value in identifying the cause of gastroenteritis. However, in parasitic infections, the main method of diagnosis is the demonstration, in unstained wet-preparations of faeces, of the ova of worms and the vegetative and cyst forms of protozoa.

   Immunofluorescence: demonstration of the causal organism by direct or indirect immunofluorescence not only detects its presence in the material under examination but also identifies it serologically.

2. *Isolation on culture* of the infecting organism: the most widely used and best method of diagnosis. It enables a complete and accurate diagnosis to be made relatively quickly and can also determine accurately the antibiotic sensitivity of the causal bacterium. The result of these tests may be essential if the patient is to receive appropriate treatment.

3. *Serology*—the demonstration of antibody to the causal organism: in general, less satisfactory than isolation. However, of great value in a few infections (e.g. Legionnaires' disease, leptospirosis, syphilis) where the organism responsible is very difficult or impossible to culture.

28

Respiratory tract infections

A very important cause of sickness — reckoned to account for a half of general-practitioner consultations and a quarter of all absences from work due to illness.

Route of infection — inhalation. Control of infections acquired by inhalation of infected secretions is well-nigh impossible. More frequent in winter time (October to March): close contact in school, at work and socially allows ready transfer of the causal agents — 'coughs and sneezes spread diseases'. In family outbreaks infection is often introduced by the most susceptible member, usually a pre-school or school-age child. Immunisation is only available against a few specific infections, e.g. diphtheria, whooping-cough, influenza.

The same clinical syndrome may be produced by a variety of agents and the same aetiological agent may produce a variety of clinical syndromes.

Respiratory infections can be classified into four groups:
1. Infections of throat and pharynx
2. Infections of middle ear and sinuses
3. Infections of trachea and bronchi
4. Infections of the lungs

1. INFECTIONS OF THROAT AND PHARYNX

CLINICAL FEATURES

Sore throat is the commonest symptom accompanied by a variable degree of constitutional upset. Typical throat appearances for the different aetiological agents are described but it is often impossible to decide on the cause of a sore throat by clinical examination alone. Over two-thirds of these infections are caused by viruses — often with sore throat as part of the common-cold syndrome: the remainder are bacterial in origin almost all due to *Streptococcus pyogenes*.

Streptococcal sore throat

Mild redness of tonsils and pharynx may be the only sign but the classical picture is of injection and oedema involving the fauces and soft palate with exudate— *acute follicular tonsillitis.*

In severe cases this may be complicated by a peritonsillar abscess (*quinsy throat*) and extension of the infection to involve the sinuses and middle ear producing *sinusitis* and *otitis media.* Systemic illness with fever is the rule and the cervical lymph nodes may be enlarged. *Scarlet fever* is a streptococcal infection—usually a sore throat—accompanied by an erythematous rash when the infecting strain of *Strep. pyogenes* produces erythrogenic toxin in a susceptible (i.e. non-immune) patient—usually a child.

Incubation period: 1-3 days.

Source: cases or carriers. After an acute attack transient carriage for a few weeks is common. Throat carriers outnumber nasal carriers but the latter, who often have an associated sinusitis, are much more effective disseminators.

Treatment: penicillin is the drug of choice. Therapy should be started parenterally and continued orally for 10 days to prevent complications and further spread of the organism to contacts. Patients hypersensitive to penicillin are given erythromycin: tetracycline resistance is common (about 20 per cent of strains) and tetracycline is therefore contraindicated.

Late complications of streptococcal infections

Streptococcal infections, usually sore throat, are sometimes followed by disease which appears to be immunologically induced. The disease is of two main kinds:
1. Rheumatic fever
2. Acute glomerulonephritis

Rheumatic fever

Clinically: the acute onset of fever, pain and swelling of the joints and pancarditis, on average 2 to 3, but up to 5, weeks after streptococcal sore throat. The most serious manifestation is involvement of the heart: patients commonly have myocarditis, sometimes in addition pericarditis and endocarditis. The disease has been said 'to lick the joints but bite the heart'.

Now relatively uncommon it was formerly a disease associated with poor living conditions and overcrowding—circumstances that facilitated the spread of streptococcal sore throat. It remains a major problem in developing countries.

Prognosis. Rheumatic fever usually clears up spontaneously although it has a marked tendency to recur: the main problem is that after the acute phase of the disease, patients later—often much later—develop as the end stage of endocardial involvement, chronic valvular disease of the heart, usually stenosis or incompetence of the mitral or aortic valves.

Pathology. Rheumatic fever is a disease of connective tissue which is almost certainly immunological in origin. The typical lesion is the *Aschoff nodule*—a pale hyaline focus with lymphocytic and macrophage infiltration and sometimes giant cells.

Immunology. Rheumatic fever may be the result of antibodies produced against protein and polysaccharide cell wall antigens of *Strep. pyogenes* cross-reacting with myocardial and heart valve tissue.

Laboratory diagnosis

The diagnosis can normally be made clinically but it is useful to check for the continuing presence of *Strep. pyogenes* in the throat or serological evidence of recent infection.

Isolation

    Specimen: throat swab.
    Culture: blood agar aerobically and anaerobically.
    Observe: for ß-haemolytic small dry colonies.
    Identify: by Lancefield grouping.

    Note. Rheumatic fever may follow infection with any serotype (Griffith type) of *Strep. pyogenes.*

Serology

    Specimen: clotted blood: paired samples should be sent.
    Examine: for antibody to streptolysin O (ASO)—a haemolysin produced by *Strep. pyogenes:* although an ASO titre of 200 units or more is regarded as significant and sera from about 75 per cent of patients with rheumatic fever give a positive result; evidence of *recent* infection requires the demonstration of rising or falling titres in two samples taken a few weeks apart. Tests are available to detect antibody

to other streptococcal products, e.g. hyaluronidase, DNAase B: they may be positive when the ASO titre is not raised. By using one or more of these serological tests it is normally possible to indicate that there has been an antecedent streptococcal infection.

Use of antibiotics

Antibiotics are required only to eradicate *Strep. pyogenes* from the throat: prophylactically on a long-term basis they are mandatory to prevent reinfection with risk of precipitating a recurrence: penicillin is the drug of choice.

Acute glomerulonephritis

Also an immunological complication which may follow streptococcal throat infection or, occasionally, impetigo. Unlike rheumatic fever, it is particularly liable to follow infection with certain serotypes of *Strep. pyogenes* notably, in throat infections, type 12 (the main nephritogenic streptococcus) and, in skin infections, type 49. The latent period between infection and symptoms is shorter than in rheumatic fever.

Clinically: acute glomerulonephritis presents 1-3 weeks after a streptococcal throat infection with haematuria, albuminuria and oedema. The oedema affects the face on waking causing a characteristic puffy appearance; as the day wears on this disappears and oedema of feet and ankles develops: oliguria is common and there may be hypertension.

Prognosis: good: the disease usually clears up spontaneously: however, it may cause permanent kidney damage and eventually progress to renal failure. Second attacks are uncommon.

Pathogenesis. Pathologically there is increased cellularity of the glomeruli with larger deposits on the outer and smaller deposits on the inner surfaces of the basement membrane. The disease is the result of an immunological process but the exact pathogenesis is unclear. The deposits are immune complexes thought to be formed by combination of antistreptococcal antibody with either: (i) streptococcal antigens already on the basement membrane or (ii) streptococcal antigens circulating in the blood to be later deposited on the membrane after complex formation. The complexes activate complement, with release of toxic substances which provoke an inflammatory reaction.

Laboratory diagnosis. Diagnosis is usually on clinical grounds: attempts should be made to confirm past or present streptococcal infection as described above for rheumatic fever.

Although less common than throat infection as a precipitating cause of the disease, streptococcal impetigo has caused some outbreaks of glomerulonephritis in the USA and in some tropical countries. It should be diagnosed by culture of a swab or sample of pus from the skin lesion.

Complement estimations. The level of C3 in serum is reduced: this has been interpreted as evidence of immune complex formation.

Use of antibiotics. Eradicate *Strep. pyogenes* if the organism is still present in the throat: penicillin is the drug of choice.

Diphtheria

Cause: Corynebacterium diphtheriae.

A severe disease: the inflamed fauces are covered with patches of sero-cellular exudate which forms a grey-white membrane. The severity is related to the extent of the membrane and the type of infecting strain of *C. diphtheriae.* Nasal diphtheria is often mild: laryngeal diphtheria is serious because of the risk of respiratory obstruction. *C. diphtheriae* produces a potent *exotoxin* which is *cardiotoxic* and fatalities from the disease are usually due to heart failure; it is also *neurotoxic* and can cause cranial and peripheral nerve palsies.

Incubation period: 2-5 days.

Source. Diphtheria is now very rare in developed countries due to effective active immunisation in childhood. However, small outbreaks continue to occur: most of those affected are unimmunised or have had incomplete courses of toxoid. The source of infection is the throat and (especially) the nose of symptomless carriers and of cases in which the disease is mild or inapparent. Convalescent carriage rarely lasts more than a few weeks.

Treatment. Antitoxin must be injected *without delay*—i.e. without waiting for bacteriological confirmation. Tracheostomy may be necessary to relieve respiratory obstruction. Penicillin or erythromycin should be given since antibiotics help to eliminate *C. diphtheriae* from the throat and, therefore, prevent further toxin production. Erythromycin is the preferred drug for the treatment of persistent carriers and for chemoprophylaxis in the unimmunised contacts of cases.

Candidosis

Oral thrush due to the yeast *Candida albicans* presents as white patches superimposed on red, raw mucous membrane which may involve the throat as well as the more common site of the mouth (Plate 15).

Source: endogenous; although candidosis may be precipitated by antibiotic treatment the patient is often debilitated by disease, e.g. malignancy and especially leukaemia.

Treatment: locally applied nystatin or amphotericin B.

Vincent's angina

An ulcerative tonsillitis which causes much tissue necrosis: often an extension of similar disease of gums and mouth (gingivostomatitis).

Source: endogenous. The causal organisms are a spirochaete *(Borrelia vincenti)* and Gram-negative anaerobic bacilli *(Fusobacterium species)* (Plate 12) found in small numbers in the normal mouth. Overgrowth to produce disease is precipitated by dental caries or poor oral hygiene, nutritional deficiency, leucopaenia (e.g. in leukaemia) and viral infections (e.g. herpes simplex, infectious mononucleosis).

Treatment: penicillin and/or metronidazole.

Infectious mononucleosis

Exudative tonsillitis is often the presenting feature of this generalised viral infection due to Epstein-Barr virus — the so-called anginose form of infectious mononucleosis.

Source: oropharyngeal secretions — the kissing disease of young adults.

Treatment: No specific treatment. Antibiotics should be avoided: almost all patients given ampicillin develop a skin rash.

DIAGNOSIS OF THROAT AND PHARYNGEAL INFECTIONS

Isolation—or *demonstration*—of the causal bacterium.

Specimen: a well-taken throat swab. Illumination of the throat and depression of the tongue are essential. The swab should be gently rubbed over the affected area so that it collects a sample of any exudate present.

Gram-stained film: a mixed bacterial flora is always present and the only findings of value are the recognition of Vincent's organisms (Plate 12) and yeasts (Plate 15).

Note. This is the only method of diagnosing Vincent's infection—the causal organisms cannot be isolated by routine culture methods.

Culture: the swab is inoculated onto a variety of media incubated at 37 °C for 24-48 h:

1. Blood agar
2. Crystal violet   —selective for *Strep. pyogenes*
    blood agar   especially if incubated anaerobically
3. Sabouraud's medium   —selective for *C. albicans*
4. Loeffler's serum slope   ⎤ —for the isolation of *C. diphtheriae*
    Blood tellurite medium ⎦ not routine nowadays

2. INFECTIONS OF MIDDLE EAR AND SINUSES

Acute infection of the middle ear or sinuses is often due to secondary bacterial invasion following a viral infection of the respiratory tract: this may be a common cold or measles—of which otitis media is a frequent complication.

Acute infections

Otitis media: an upper respiratory infection involving the middle ear by extension of infection up the Eustachian tube. Predominantly a disease of children: the main symptom is *earache.*

On examination the eardrum is injected or red and the infection may progress to cause bulging with eventual rupture of the tympanic membrane and discharge of pus from the ear. Recurrent attacks are common.

Sinusitis: mild discomfort over the frontal or maxillary sinuses due to congestion is a frequent symptom in common colds: severe pain and tenderness with purulent nasal discharge, however, indicate bacterial infection and require treatment.

Causal bacteria: Haemophilus influenzae, Streptococcus pyogenes, Streptococcus pneumoniae.

Source: endogenous spread of organisms from the normal flora of the nasopharynx.

Diagnosis. In the majority of cases of sinusitis and in many of otitis media, specimens from the site of infection cannot be obtained. If the eardrum ruptures or if myringotomy (incision of the tympanic membrane to release pus in the middle ear) is performed, collect a swab of exudate; if drainage or lavage of the sinuses is carried out, material should be collected and cultured in the same way as a sample of pus on a range of suitable media.

Treatment: amoxycillin or ampicillin, alternatively erythromycin.

Chronic infections

Chronic suppurative atitis media. Characterised by suppuration in the middle ear distorted by chronic pathological changes. Periods of quiescence when the ear is relatively dry are followed by exacerbations when there is profuse discharge associated with pain. This is usually a long-standing disease which can recur at intervals throughout childhood and into adult life.

Chronic sinusitis. Painful sinuses with headache are prominent symptoms; often associated with nasal obstruction with mucoid or purulent nasal discharge.

Causal bacteria: are the same as those implicated in acute infections—i.e. infection is usually endogenous with bacteria from the normal upper respiratory flora. A variety of other organisms may be found, including *Staph. aureus* and a wide range of coliform bacilli. Pseudomonads and proteus are common in chronic ear discharges and also bacteroides; detection of these anaerobes requires the careful laboratory examination of a well-taken specimen. The clinical significance of some of these organisms is uncertain.

Diagnosis. Swabs of pus from the ear; lavage specimens from the sinuses—such saline washings are always contaminated by nasal flora. Examine as specimens of pus.

Treatment of chronic infections. Antibiotics often give disappointing results: if prescribed therapy should be guided by antibiotic sensitivities of isolated organisms but treatment may have to be on a 'best-guess' basis. Topical antimicrobials (e.g. neomycin or framycetin, polymyxin, bacitracin) are given in chronic otitis media since systemic drugs fail to penetrate to site of infection but there is little evidence that they are effective.

3. INFECTIONS OF TRACHEA AND BRONCHI

Laryngitis, tracheitis and bronchitis are usually associated with or follow a viral infection of the upper respiratory tract.

Laryngitis

Clinical features: hoarseness and loss of voice: in more severe form, *croup* (or acute laryngotracheobronchitis) with croaking cough and stridor. In children, most often associated with parainfluenza virus infection: occasionally due to a rare but important bacterial infection—acute epiglottitis.

Acute epiglottitis

Clinical features: severe croup syndrome in children (usually aged between 2 and 7 years) which may rapidly progress to respiratory obstruction and death. The epiglottis is inflamed and oedematous.

Causal bacterium: capsulated strains of *Haemophilus influenzae* (usually of type b).

Diagnosis: H. influenzae may be isolated from the epiglottis and from blood culture.

Treatment: parenteral amoxycillin or ampicillin: tracheostomy may be necessary.

Bronchitis

Clinical features: a feeling of tightness in the 'tubes', cough, initially dry and painful, later productive with expectoration of yellow-green sputum most marked in early morning specimens; variable degree of fever and of constitutional upset. Abnormal chest signs, e.g. rhonchi are found on auscultation.

Acute bronchitis in a patient with a healthy respiratory tract is often a trivial complication of a viral upper tract infection: the initial viral attack damages respiratory mucous membrane with paralysis of ciliary movement. Although viral acute bronchitis is usually mild and self-limiting, secondary bacterial infection often supervenes in more severe attacks especially in patients with chronic respiratory disease such as chronic bronchitis, asthma and bronchiectasis.

Chronic bronchitis: acute exacerbations of chronic bronchitis are serious events in the course of a major killing disease. Chronic bronchitis is not itself due to infection; aetiological factors include low socio-economic class, urban dwelling (atmospheric pollution) and tobacco consumption, especially cigarette addiction ('smoker's cough'). Exacerbations, however, are associated with bacterial infection. They commonly follow respiratory infections or a fall in atmospheric temperature with increase in humidity (together causing foggy weather): all these factors are often present concurrently in winter. During exacerbations both volume and purulence of sputum increase.

Pathological changes in chronic bronchitis are: (i) increase in the number of mucous-containing cells in the bronchi with consequent hypersecretion of mucus; (ii) inflammation, fibrosis, collapse, dilatation and cyst formation in bronchioles and alveoli. After exacerbations some changes may resolve but others do not, resulting in progressive, irreversible damage.

Causal bacteria and source:

1. *Haemophilus influenzae* (usually non-encapsulated strains)—the closely related organism—*H. parainfluenzae* is sometimes isolated but is of doubtful pathogenicity.
2. *Streptococcus pneumoniae* (pneumococcus).
3. *Branhamella catarrhalis:* disregarded for decades as a potential cause of respiratory infection but recent reports have drawn attention to its undoubted pathogenic role in some exacerbations of bronchitis.

   All three organisms are together present in the upper respiratory tract in health. The secondary bacterial invaders in bronchitis are therefore endogenous. Normal subjects have a sterile bronchial tree but in chronic bronchitis the bronchi become colonised, especially with *H. influenzae*, even when the disease is quiescent. During exacerbations the concentration of *H. influenzae* in respiratory secretions increases along with sputum purulence. Specific antibodies to *H. influenzae*, absent in non-smoking healthy adults, are present in the serum of two-thirds of chronic bronchitics. *H. influenzae* is now regarded as the prime pathogen in exacerbations of chronic bronchitis.
4. *Mycoplasma pneumoniae:* although usually associated with pneumonia, this agent causes a wide spectrum of respiratory disease and is often unrecognised as an aetiological agent in bronchitis. Patients are typically school-age children and young adults. Source is exogenous and spread is by the respiratory route. Cases are usually sporadic but there may be family outbreaks and, occasionally, institutional epidemics. The frequency of infection in the community varies from year to year.

Treatment—acute bronchitis

1. In previously healthy subjects, acute bronchitis usually subsides in 2-5 days and does not require antibiotic therapy.
2. The majority of patients with troublesome, purulent, acute bronchitis are given antibiotics prescribed on an informed best-guess basis. The drugs must be active against both *H. influenzae* and pneumococci: fortunately both of these bacteria have predictable sensitivity patterns—although there are some β-lactamase producing *H. influenzae* which are resistant to ampicillin/amoxycillin and also some pneumococci which are resistant to tetracycline. About half of the strains of *B. catarrhalis* produce a β-lactamase. Laboratory examination of sputum is essential if the patient fails to respond to an apparently adequate course of treatment.

Treatment—chronic bronchitis

Treatment should start as early as possible and chronic bronchitics should have an emergency supply of antibiotics to take whenever a cold 'goes to the chest': this shortens the duration and reduces the severity of exacerbations but, unfortunately, does not prevent deterioration of respiratory function.

Short-term treatment (5-7 days)

1. *Ampicillin or amoxycillin:* may be bactericidal in action in adequate dosage. Amoxycillin is preferred because of better absorption in the presence of food and ability to penetrate mucoid sputum: ampicillin only attains bactericidal levels against *H. influenzae* when the sputum is purulent. *Augmentin* may be indicated when the infecting strain is shown to be a β-lactamase producer.
2. *Tetracyclines:* bacteriostatic action, uncertain sputum penetration. An advantage is the activity against *M. pneumoniae*.
3. *Cotrimoxazole:* bacteriostatic action: the ratio of sulphonamide to trimethoprim in the sputum is 4 to 1 because of poor sulphonamide penetration (compared to synergistic bactericidal ratio of 20 to 1 found in blood). *Trimethoprin* alone may be prescribed as an alternative.
4. *Chloramphenicol:* a reserve drug because of bone-marrow toxicity. Although bacteriostatic it is very effective due to excellent penetration and high activity against *H. influenzae*.

Long-term treatment

Winter-long suppressive tetracycline treatment for advanced chronic bronchitis was claimed to reduce the number and severity of exacerbations. Fashionable 20 years ago, this continuous bacteriostatic regimen has been abandoned in favour of short-term courses of bactericidal treatment.

Vaccines

Viral: the only vaccine available is that against influenza. Since chronic bronchitics often suffer severe exacerbations after influenza and are at risk of developing secondary bacterial pneumonia, it is recommended that they be vaccinated annually. The vaccine contains inactivated virus of currently circulating A and B strains and achieves protection of the order of 70 per cent.

Bacterial: polyvalent pneumococcal polysaccharide vaccines are now available. They may be of value in preventing pneumococcal pneumonia in this 'at risk' group.

Diagnosis

See 'Diagnosis of bacterial chest infections' on p. 179.

Cystic fibrosis

Due to improved management more infants and children with this disease, transmitted as an autosomal recessive trait, survive to adult life.

This inherited defect leads to production of abnormally viscid mucus which blocks tubular structures in many different organs: the most disabling obstructive changes affect the lungs and chronic respiratory infection is a major problem.

Causal bacteria:

1. *Staphylococcus aureus* initially, tending to be replaced by:
2. *Pseudomonas aeruginosa*

Treatment: appropriate antistaphylococcal or antipseudomonal drugs alone or in combination, the choice dependent on in vitro antibiotic sensitivity results. Long-term administration may be required.

Pertussis (whooping-cough)

Clinical features

An acute tracheobronchitis of childhood.

Onset is insidious—initially a catarrhal stage with common-cold symptoms which lasts about 2 weeks: followed by a stage of paroxysmal coughing (2 weeks): residual cough persisting for a month or more is a common sequel.

Paroxysmal cough is a diagnostic feature: it consists of repeated violent exhalations with a distressing, severe inspiratory whoop: there is expulsion of tenacious clear bronchial mucus: vomiting is common (Plate 16).

Fatality is low: but morbidity may be high: there is a significant risk of developing subsequent chronic chest disease, e.g. bronchiectasis. Most acute deaths are in infants during the first year—and especially in the first 6 months of life.

Causal bacterium: Bordetella pertussis (types 1, 3: 1, 2, 3 and 1, 2). Type 1, 3 is responsible for over half of the infections. *(Bord. parapertussis* causes mild whooping cough and is relatively rare in Britain).

A similar syndrome may be caused by adenoviruses (types 1, 2 and 5) and *M. pneumoniae.*

Epidemiology: following the decline in the acceptance of immunisation in the mid-1970s, there have been two epidemics of whooping cough in the UK, between 1977–1979 and 1981–1982.

Incubation period: 1-3 weeks: often about 10 days.

Source: patients—most infective during catarrhal stage, becoming non-infective at end of paroxysmal stage.

Spread: airborne via droplets.

Diagnosis

Isolation: isolation from infected clinical cases is not easy, and diagnosis is usually based on symptoms. Organisms are much less numerous after catarrhal stage (i.e. when typical symptoms develop) and in immunised patients.

Specimen:

1. *Pernasal swab*—passed gently along the floor of the nose to sample naso-pharyngeal secretions.
2. *Cough plate*—held in front of mouth during a paroxysm of coughing: superseded by pernasal swabbing.

Inoculate: charcoal blood agar—the preferred medium—or Bordet-Gengou medium.

Incubate: 3-5 days at 35-36 °C.

Observe: moist 'mecury-drop' colonies.

Identify: by slide agglutination with specific antisera.

Serology. Rising titres of antibody in *paired* serum specimens may be diagnostic in older children (over 1 year). Detected by complement fixation, agglutination or immunofluorescence tests. These may give evidence of infection when the opportunity for bacterial isolation has been missed. Interpretation of results is especially difficult in fully or partially immunised children.

Treatment: antibiotics are only of value if given within the first 10 days of infection i.e. during the catarrhal stage when the diagnosis may not be suspected. If secondary pneumonia develops it should be treated appropriately (depending on the antibiotic sensitivity of the organisms isolated).

Administration of ampicillin or erythromycin to the patient reduces the duration of infectivity and these drugs are successful chemoprophylactics when given to close contacts (e.g. siblings).

Vaccination: this controversial topic is discussed on page 353.

4. INFECTIONS OF THE LUNGS

Pneumonia

The most severe and life-threatening of respiratory infections: although antibiotic therapy has transformed the prognosis for many patients, pneumonia remains a significant cause of death—in infancy, in the elderly and in immunocompromised patients.

Clinical features

Onset: sometimes abrupt, sometimes insidious when due to the extension of a pre-existing respiratory infection.

Symptoms: fever, rigors, malaise: respiratory symptoms include shortness of breath, rapid shallow breathing, cyanosis, cough, sometimes pleural pain: sputum may be initially tenacious and rusty—later becoming purulent.

Clinical investigation: signs of consolidation of lungs i.e. dullness on percussion, reduced air entry, moist rales: there is usually polymorphonuclear leucocytosis.

Radiology: indicates the site and extent of the consolidation.

Pneumonia is of three main types:
1. *Lobar (or segmental) pneumonia:* in which the consolidation is limited at least initially to one lobe or segment of the lung: the main type of pneumonia seen in previously healthy people.
2. *Bronchopneumonia:* usually bilateral the consolidation is scattered throughout the lung fields although it is mainly concentrated at the bases; the most common form of pneumonia—seen principally in the elderly and in patients with debilitating or chronic respiratory disease such as chronic bronchitis, bronchiectasis.
3. *Primary atypical or virus pneumonia:* patchy consolidation of the lungs with widespread opacities—often out of proportion to the mild degree of illness and the few clinical signs: usually a self-limiting disease.

Legionnaires' disease. An extremely severe pneumonia first recognised in 1976 amongst members attending an American Legion

Convention.

Table 28.1 lists the main organisms associated with the different types of pneumonia.

Table 28.1 Causes of pneumonia

Pneumonia	Main causal organisms
Lobar	Streptococcus pneumoniae
Bronchopneumonia	Streptococcus pneumoniae
	Haemophilus influenzae
	(rarely Staphylococcus aureus, coliforms)
Primary atypical pneumonia*	Mycoplasma pneumoniae
	Coxiella burneti
	Chlamydia psittaci
Legionnaires' disease*	Legionella pneumophila

*these are multisystem diseases which affect other organs as well as the lungs.

CAUSAL AGENTS

1. **Streptococcus pneumoniae** (pneumococcus): the main cause of pneumonia.

   Source: human respiratory tract.

   Spread:

   a. *Exogenous droplet transmission of virulent strains* (incubation period 1-3 days)

   b. *Endogenous* due to downward spread of pneumococci from the flora of the nasopharynx

   Treatment. Recommended drug: penicillin. Other effective drugs: ampicillin/amoxycillin, cephalosporins, erythromycin, cotrimoxazole.

2. **Haemophilus influenzae:** often underestimated as a cause of pneumonia; in infants usually due to capsulated strains: in adults, often but not always, complicating chronic respiratory disease *(note,* in pneumonia following exacerbations of chronic bronchitis the pneumococcus is the commonest bacterial cause).

   Treatment. Recommended drug: ampicillin/amoxycillin: augmentin. Other effective drugs: tetracycline, cotrimoxazole.

3. **Staphylococcus aureus:** a relatively uncommon cause of bronchopneumonia: probably most often seen in hospital patients: sometimes after influenza the cause of severe secondary bacterial pneumonia: an increasing problem in intravenous drug abusers: often fulminating and rapidly fatal.

Treatment. Recommended drug: cloxacillin (penicillin if the infecting strain is sensitive). Other effective drugs: fusidic acid, lincomycins, erythromycin, vancomycin.

4. **Coliforms** e.g. *Escherichia coli, Proteus species, Klebsiella species, Pseudomonas species:* rare causes of bronchopneumonia: their isolation from sputum often indicates merely colonization of the respiratory tract, e.g. after a course of antibiotic, and must be interpreted cautiously: as a cause of pneumonia they are most often encountered in hospital patients especially the immunocompromised (e.g. patients with leukaemia) or those on support ventilation under intensive care.
   Treatment. Recommended drugs: an aminoglycoside often in combination with an appropriate β-lactam drug. Review regimen when results of laboratory sensitivity tests available.
   Friedlander's bacillus: a klebsiella of uncertain taxonomy; has been described as a cause of a rare pneumonia with much tissue destruction.

5. **Mycoplasma pneumoniae** (formerly known as Eaton agent): the main cause of *primary atypical or 'virus' pneumonia:* school-age children and young adults are the group most affected.
   Incubation period: about 3 weeks.
   Spread: respiratory (droplet) spread to involve individuals, families and sometimes institutions (e.g. schools, military camps): the prevalence peaks every 3-5 years.
   Treatment. Recommended drug: tetracycline. Other effective drug: erythromycin.

6. **Coxiella burneti:** responsible for the acute febrile disease Q fever; up to half of the patients have pneumonia—usually a patchy consolidation.
   Treatment. Recommended drug: tetracycline. Other effective drug: erythromycin.

7. **Chlamydia psittaci:** the causal agent of ornithosis and psittacosis in birds: may infect man via inhalation of dried bird droppings: produces an acute influenza-like illness with patchy pneumonia.
   Treatment. Recommended drug: tetracycline. Other effective drug: erythromycin.

8. **Legionella pneumophila and related genera:** cause of Legionnaires' disease; now being increasingly recognised as a cause of pneumonia. Patients are typically middle-aged smokers often in poor general health.
   Symptoms: initially influenza-like, the illness may progress to a severe pneumonia sometimes with respiratory failure; other promi-

nent features are mental confusion, acute renal failure and gastrointestinal symptoms.

Source: environment: usually associated with water.

Spread: by contaminated aerosols, e.g. via air-conditioning systems, water from storage tanks through showers and taps, etc. Person-to-person droplet spread has not been recorded.

Treatment. Recommended drug: erythromycin. Other effective drug: rifampicin. These may be given together.

VIRUSES

Pneumonia in infants is predominantly due to respiratory syncytial virus: primary influenzal pneumonia in previously healthy adults is a rare but exceedingly severe complication of influenza and is almost invariably fatal: however, viruses on their own are uncommon causes of pneumonia except in immunocompromised patients and antibiotic treatment is therefore indicated for all patients diagnosed as having pneumonia.

TREATMENT OF PNEUMONIA

This is governed by the clinician's experience, his personal preference and also knowledge of what the infecting agent is likely to be. As a rule, treatment has to be started before laboratory results are available and even after investigation, the cause of pneumonia in some patients is never identified.

The penicillins are the drugs of first choice: in patients with lobar pneumonia, when the infecting organism is likely to be *Strep. pneumoniae* alone, prescribe penicillin; in bronchopneumonia, give ampicillin or amoxycillin since *H. influenzae* may also be involved.

In patients who either present with a *severe* undiagnosed pneumonia or fail to respond to initial therapy, ampicillin and erythromycin should be given in combination. Sometimes treatment will be changed if laboratory investigation indicates a more appropriate antibiotic.

Tuberculosis

Although tuberculosis is not considered in this chapter (see Chapter 37) in any long-standing chest infection, the **possibility of tuberculosis must always be considered.**

Aspiration pneumonia, lung abscess

Aspiration pneumonia follows inhalation of vomit or sometimes a foreign body in an unconscious patient: the causal organisms are commensals of the upper respiratory tract—principally *Strep. pneumoniae* but anaerobic bacteria notably *Fusobacterium species, Bacteroides melaninogenicus* and also anaerobic cocci are involved in the majority of cases.

Lung abscess is nowadays rare: usually due to obstruction of a bronchus or bronchiole, e.g. by an inhaled foreign body or to suppuration developing within an area of pneumonic consolidation: the infecting organisms are similar to those listed above—namely *Strep. pneumoniae* and anaerobic bacteria of the upper respiratory flora.

Anaerobic organisms are seldom isolated from sputum. Samples should be collected through a fine catheter which is usually passed down a bronchoscope but sometimes introduced by the transtracheal route through the cricothyroid membrane: an alternative method is transthoracic needle aspiration.

Treatment. Recommended drugs: penicillin and metronidazole.

Empyema

Literally, pus in the pleural space and nowadays a rare complication of pneumonia: sometimes tuberculous: laboratory diagnosis requires aspiration and treatment depends on drainage and removal of the infected fluid with appropriate antibiotic therapy: usually due to *Strep. pneumoniae, Staph. aureus,* with upper respiratory anaerobes sometimes implicated.

DIAGNOSIS OF BACTERIAL CHEST INFECTIONS

1. *Isolation* of causal pathogen from sputum or, less commonly, from the aspirate of a pleural effusion, lung abscess or area of pneumonic lung. *Blood cultures* should also be taken since the infecting bacterium may be present in the blood of up to one third of patients with pneumonia.
2. *Detection of bacterial antigen in sputum.*
3. *Serological:* demonstration of specific antibody in patient's serum is only useful in cases of atypical pneumonia (see below).

Sputum examination and culture

Specimen: early morning sputum since this sample is likely to be the most purulent.

Collection: try to minimise salivary contamination—but the presence of some oro-pharyngeal flora is inevitable.

Transport: send to laboratory without delay: in transit, delicate organisms (e.g. *H. influenzae*) die, robust bacteria (e.g. coliforms) multiply and overgrow.

Laboratory examination

Macroscopic—note naked-eye appearance.
Microscopic:

1. *Gram film:* observe amount of pus, squamous epithelial cells (indicating buccal contamination) and nature of the bacterial flora—if this is very mixed it is probably not significant but the presence of a predominant organism may allow an immediate provisional diagnosis. Plate 4 shows a typical Gram film of sputum from a patient infected with both pneumococci and *H. influenzae.*
2. *Ziehl-Neelsen or auramine film:* examine for acid and alcohol fast bacilli; if present, a presumptive diagnosis of tuberculosis can be made. This examination is not always carried out nowadays because of the decline in the incidence of tuberculosis—but should always be done in immigrants or in areas where tuberculosis is still common (e.g. large cities) or if there is any clinical indication of tuberculosis.

Culture: bacteria are not distributed evenly throughout sputum; select a purulent portion: alternatively homogenise the specimen by treatment with a liquifying agent: such treatment allows semiquantitative culture and may make evaluation of results easier.

Inoculate:

1. Blood agar: observe for presence of a predominant organism and assess respiratory flora.
2. Chocolate agar with bacitracin: selective for *H. influenzae.*
   Incubate 1. and 2. at 37 °C in air with 5-10 per cent carbon dioxide—this atmosphere is necessary for the primary isolation of some strains of *H. influenzae* and *Strep. pneumoniae.*
3. Blood agar incubated at 37 °C anaerobically with 5-10 per cent carbon dioxide: this is not done routinely although *Strep. pneumoniae* grows better under these conditions: indicated if infection with non-sporing anaerobes such as bacteroides is suspected e.g. in lung abscess or aspiration pneumonia.

Assessment of culture results may be difficult because of contamination from the oropharyngeal flora. Viridans streptococci, neisseriae, coagulase-negative staphylococci, commensal corynebacteria, etc. are normally regarded as upper respiratory tract commensals. Small numbers of haemophilus and pneumococci in a mixed growth may be part of the normal flora; larger numbers, especially if other bacteria are scanty, are regarded as pathogens.

The oropharynx of patients who have received antibiotics often becomes colonised with coliform organisms and yeasts and, under such circumstances, undue significance should not be placed on their isolation.

Culture for Mycobacterium tuberculosis: not a routine examination nowadays but should be carried out if indicated clinically. Submit three samples collected on successive days for culture on Lowenstein-Jensen medium for 6 to 8 weeks.

Detection of bacterial antigen in sputum:

Pneumococcal capsular antigens can be demonstrated by a variety of methods. A positive result may be obtained when culture of the same specimen failed to isolate *Strep. pneumoniae* usually because the patient had received antibiotics before the sputum was collected. In some patients with Legionnaires' disease, *L. pneumophila* antigen can be detected.

Serological tests

The atypical pneumonias and Legionnaires' disease are difficult to diagnose by isolation of the causal organism. Suspected cases should be investigated serologically i.e. by detection of antibody in patient's serum: most often diagnosed by stationary high titres (although demonstration of a four-fold rising titre is better).

The tests used to diagnose these diseases are shown in Table 28.2.

Table 28.2 Serological tests in atypical pneumonias

Causal organism	Test used to detect antibody	Diagnostic (stationary) titre
Mycoplasma pneumoniae	Complement fixation	256
Coxiella burneti	Complement fixation	256
Chlamydia psittaci	Complement fixation	256
Legionella pneumophila	Indirect immuno-fluorescence	256

29

Diarrhoeal diseases

Most but not all cases of diarrhoea are due to infection: bacteria, viruses and protozoa cause diarrhoea but this chapter will deal principally with bacteria.

Host: the young are most susceptible; poor general health and nutrition also predispose to diarrhoeal disease.

Bacteria: factors such as size of infecting dose, enterotoxin production, ability to adhere to gastrointestinal epithelium and to invade the gut wall affect the ability of organisms to infect the gut.

Epidemiology: prevention of spread of diarrhoeal disease in a community largely depends on sanitation (i.e. adequate disposal of sewage), clean food and a safe water supply; personal hygiene (i.e. washing hands after defaecation) is a remarkably effective means of preventing faecal-oral spread.

Storage of food at room temperature must be avoided: this permits rapid bacteria multiplication and, with some bacteria, the formation of toxins. Note that although *refrigeration* prevents bacterial multiplication, the bacteria are preserved, not killed, at 4 °C.

INVESTIGATION OF DIARRHOEA

1. *History:* particularly with regard to
   a. Clinical symptoms, duration, etc.
   b. Recent foreign travel.
   c. Food history including symptoms amongst other consumers of suspect food and probable incubation period.
2. *Specimens for laboratory examination:*
   a. Faeces (rectal swab if none available).
   b. The suspected food: strenuous efforts should be made to obtain samples: this is often difficult as it has usually been eaten or discarded.
   c. Vomit.

Table 29.1 Causal organisms and some features of diarrhoeal diseases

Organism	Usual source	Common mode of spread
Salmonella species	Animal gut	Poultry; meat; milk
Campylobacter species	Animal gut	Poultry; meat; milk
Shigella species	Human gut	Faecal-oral or via food, fomites
Escherichia coli	Human gut	Faecal-oral or via food, water, fomites
Staphylococcus aureus	Septic lesions on food handlers	Cooked meats, dairy products
Clostridium perfringens	Animal gut	Stews; meat pies
Bacillus cereus	Environment (soil)	Rice
Clostridium difficile	Human gut	Probably faecal-oral
Yersinia enterocolitica *Yersinia pseudotuberculosis*	Animal gut	Faecal-oral, via food, water
Vibrio cholerae	Human gut	Water and food

    d. Blood culture: in severe cases—especially the very young and the elderly.

Other investigations which are especially important in an outbreak include

1. Food-handling practices in the kitchen concerned.
2. Faecal samples from kitchen staff.

The main enteropathogenic bacteria with some characteristics of the diseases they cause are shown in Table 29.1. It is clear that there are many different bacteria which can cause diarrhoea.

SECTION I: COMMON BACTERIAL CAUSES OF DIARRHOEAL DISEASES IN BRITAIN

The main bacterial causes of diarrhoea in this country are:
Salmonella
Campylobacter
Shigella
E. coli
Staph. aureus
Cl. perfringens
The diseases these organisms cause are described below in Section I.

The rarer causes of diarrhoea in Britain are discussed in Section II: Cholera, once epidemic in Britain and still a major problem in the Third World is described in Section III.

SALMONELLA

Probably the commonest cause of diarrhoea in Britain—and the incidence is still increasing. Diarrhoea due to salmonella is, by tradition, called food-poisoning although this term is somewhat misleading.

Cause: Salmonella species; there are more than 1500 serotypes but only about 14 are important or common causes of infection.

(Note: Salmonella typhi, paratyphi A, B and *C* classically cause enteric fever, a septicaemic febrile illness in which diarrhoea is a late symptom (Chapter 9). *S. paratyphi* is intermediate in its pathogenicity and can cause either mild enteric fever or a primarily diarrhoeal illness.

Habitat: domestic animals, poultry.

Clinical features

Asymptomatic infections are common as are cases with only mild gastrointestinal disturbance.

Incubation period is short—around 12-36 hours.

Main symptoms are acute onset of abdominal pain and diarrhoea, sometimes with fever and vomiting; dehydration may require correction especially in babies.

Septicaemia: sometimes develops in severe cases and is more common with certain serotypes (e.g. *S. cholerae-suis, S. enteritidis).*

Pathogenesis

Not well understood: site of infection is either the small or large intestine; some strains produce enterotoxins similar to those of toxigenic strains of *Esch. coli:* other salmonellae invade the mucosa of the small intestine—like shigellae.

Diagnosis

Isolation

   Specimen: faeces.

   Culture: on MacConkey's (indicator) medium and (selective media) such as desoxycholate citrate agar and Wilson and Blair's medium. Selenite F is a good enrichment medium but since it is a broth, requires subculture to MacConkey medium after 24 h incubation.

Observe: for pale, non-lactose fermenting colonies (MacConkey and desoxycholate citrate media) or black shiny colonies (Wilson and Blair's medium).

Identify: initially by biochemical tests; then serologically to determine 'H' and 'O' antigens.

Treatment

Antibiotics are contraindicated except in septicaemic cases: they do not affect symptoms and may prolong convalescent carriage of the organism: they also contribute to emergence of antibiotic-resistant strains.

Treatment is rarely necessary: rehydration is occasionally required in babies.

Epidemiology

Food derived from domestic animals and poultry is the main source: this is usually meat which is contaminated from viscera at slaughter but there are occasional outbreaks due to contaminated milk. Human cases and carriers may also play a role but this seems to be surprisingly rare.

Food, especially meat and offal, is often contaminated in the raw state. If it is then inadequately cooked and stored for some time at a warm room temperature, surviving salmonellae can multiply: alternatively, salmonellae from raw meat can contaminate other cooked foods by common use of kitchen tools and work-surfaces with subsequent multiplication during storage.

Outbreaks of salmonella food-poisoning are common: they often involve communal catering, e.g. weddings, large dinners, etc., but may be a problem in hospitals especially in mental or geriatric units.

Serotypes: The most common is *Salmonella typhimurium;* other common salmonellae are *S. enteriditis, S. hadar,* and *S. virchow.*

Control

Difficult: control is clearly unsuccessful at present. It depends on:
1. Control of imported animal feedstuffs which may cause infection of domestic animal stock.

2. Good farming and abattoir practice, restricted prescribing of antibiotics especially to calves since indiscriminate use results in selection of multiple-resistant bacteria which can then spread to human populations.
3. Rigorous hygiene in the kitchen, e.g. separation of cooked and raw foods, prompt and efficient refrigeration of cooked food, thorough thawing of both frozen poultry and meat before cooking.
4. Good personal hygiene in food handlers.
5. Exclusion of known human excretors from food handling.

Note: cooking is still the best method of rendering food safe from infection before eating it.

CAMPYLOBACTER

This organism is now recognised as a major cause of diarrhoea—in fact second only to Salmonella. It requires a high temperature for growth and its frequency and importance were only appreciated when special methods of culture were used.

Campylobacters are small vibrio-like organisms: the classification of individual species is confused; cause of human infections is *C.fetus* subspecies *jejuni* but subspecies *intestinalis* (which grows at 25 °C rather than high temperatures) is occasionally involved.

Clinical features

Two clinical presentations but like most intestinal infections symptoms vary from mild (or even symptomless) to a severe illness with prostration.

1. *With prodromal symptoms:* of fever, headache, backache, limb pain, nausea and abdominal pain: after 24 h, sometimes longer, diarrhoea develops.
2. *Without prodromal symptoms:* acute onset of abdominal pain and diarrhoea.

Incubation period 3 to 10 days.

Diarrhoea: is often severe with *blood* and mucus in stools: up to 20 stools a day may be passed and there may be faecal incontinence.

Abdominal pain is a prominent feature of campylobacter infection.

Septicaemia: with fevers, rigor and malaise is sometimes seen in severe cases confirming the invasiveness of the causal organisms.

Duration is often several days and relapses are common.

Pathogenesis

Typically an enterocolitis; infection involves the ileum but is not restricted to the small intestine and there is often colitis also.

Histology: there is some suggestion from the results of gut biopsy that campylobacters are invasive and penetrate beyond the mucosa.

Diagnosis

Isolation

Specimens: faeces.

Culture: on selective medium (containing vancomycin, polymyxin and nystatin) at 43 °C (but, note, the rarer *C.fetus* subspecies *intestinalis* grows only at 25 °C.)

Observe: for typical colonies.

Identify: by morphology—curved, slender Gram-negative bacilli, with characteristic darting motility, positive oxidase reaction and other features.

Treatment

Usually self-limiting: erythromycin has been used to relieve symptoms but recent work suggests that it is not effective.

Epidemiology

Source of infection: farm animals—especially poultry—are probably the major source of human infection: milk has been incriminated; dogs and cats have also been reported as sources of campylobacters.

Route of infection: eating contaminated food; but there can be faecal-oral spread especially between children.

Control

Food and personal hygiene: control of infection in animals is not practicable.

SHIGELLA

Shigellae cause bacillary dysentery—'the commonest of the unpreventable diseases'. A worldwide problem and an important cause of death and morbidity in young children, especially in the Third World.

Cause: Shigella—the four species are as follows:
1. *Sh.dysenteriae* (formerly *Sh.shiga*): 10 serotypes.
2. *Sh.flexneri:* six serotypes.
3. *Sh.sonnei:* one serotype (the main cause of dysentery in Britain).
4. *Sh.boydi:* 14 serotypes.

Clinical features

Incubation period: from 1–9 days.

Symptoms: diarrhoea with blood, mucus and often pus in the stools which varies from a severe life-threatening disease to a mild or symptomless infection.

Shiga dysentery is due to *Sh.dysenteriae:* a severe, even life-threatening disease found only in tropical countries with fever, abdominal pain and diarrhoea: the disease sometimes becomes septicaemic, indicating that, unlike other shigellae, *Sh.dysenteriae* has invasive properties. *Sh.dysenteriae* produces a powerful neurological exotoxin but this probably does not play a role in Shiga dysentery: an enterotoxin and cytotoxin are also produced—their role is uncertain but they may be partly responsible for the organism's invasiveness.

Dysentery due to other shigellae is generally a milder disease which varies from asymptomatic excretion to a prostrating attack of diarrhoea with abdominal pain and—usually minimal—fever: the stools may contain blood, mucus and pus but blood is unusual in cases of Sonne dysentery:

Shigella sonnei is the usual cause of dysentery in Britain. Sonne dysentery, however, is worldwide. The main age group affected are young children, particularly those in nursery schools where, because of the opportunities for faecal-oral spread, epidemics are not uncommon; the disease is usually mild but in a few cases dehydration ensues requiring emergency treatment; epidemics of Sonne dysentery are also frequent in mental hospitals and the infection may be difficult to eradicate.

Shigella flexneri was formerly quite common in Britain: it persisted as a major cause of dysentery in a few areas such as Glasgow long after it had disappeared from the country as a whole; improved housing conditions are thought to have been responsible for its virtual disappearance in Glasgow in recent years; still common overseas—mainly in tropical countries.

Shigella boydii is rare in Britain and the infection has almost always been acquired overseas: common in the Middle and Far East.

Diagnosis

Isolation

Specimens: stools, rectal swabs.

Culture: MacConkey (indicator) and desoxycholate citrate agar (selective) media. Selenite F enrichment broth with subculture to MacConkey's medium after 24 h incubation.

Observe for pale (non-lactose fermenting) colonies but *note: Sh. sonnei* is a late lactose fermenter and may produce pale pinkish colonies.

Identify: by biochemical tests then serologically by testing for agglutination with antiserum to *Sh. sonnei* and, if necessary, polyvalent antisera to other *Shigella species.*

Treatment

Antibiotics are rarely necessary for Sonne dysentery: the more severe forms should be treated systemically depending on the sensitivity of the organism isolated.

Epidemiology

Reservoir of infection is the human gut: not only are the symptomless infections common—with excretion of the organism in the faeces— but after an acute attack, a proportion of patients continue to excrete shigellae for some time (i.e. for weeks, sometimes months). Patients with acute dysentery, however, are the most dangerous sources of infection, doubtless due to the large numbers of shigellae excreted during the acute phase of the disease.

Route of infection is faecal-oral either directly or via contaminated equipment and towels: contamination of lavatory seats is a particularly common source of infection in nursery schools and shigellae can remain viable for long periods of time in the cool, moist environment of the average nursery school toilet.

'Food, flies, fomites' are the classical means of spread of dysentery; contaminated water can also be a source of infection.

Dysentery waxes and wanes in incidence over long periods of time: after two decades of high incidence which started in 1950, the disease waned in Britain in the late 1970s and early 1980s but is now increasing in incidence once more; this periodicity appears unrelated to improved sanitation or other public health measures.

Control

Good sanitation with safe water, adequate sewage disposal: a high standard of personal hygiene is important but almost impossible to achieve in conditions of poor housing and poverty and in nursery schools.

ESCHERICHIA COLI

Although *Escherichia coli* is part of the normal commensal gut flora, paradoxically, certain strains can be a cause of diarrhoea.

Esch. coli strains which cause diarrhoea: several mechanisms of enteric pathogenicity have been discovered in these strains. But the picture is confusing because not all the strains associated with disease have them.

1. *Enterotoxins:* these are of two types:
   (i) *LT:* heat-labile and plasmid-mediated. Detected by cytopathic effect on Y1 mouse adrenal or CHO chinese hamster ovary cell lines.
   (ii) *ST:* Heat-stable and also plasmid-coded. Detected by intestinal dilatation in suckling mice after injection of culture filtrates into stomach.
2. *Adhesive factors:* now known as colonization factor antigens (CFA) I, II and III.
   Detected by lectin-binding or mannose-resistant haemagglutination of erythrocytes: mediated by plasmid-coded pili.
3. *Enteroinvasiveness:* confers the ability to penetrate intestinal epithelial cells detected by the Sereny test in which culture filtrates produce keratoconjunctivitis after installation into the guinea-pig eye.

O Antigens

Diarrhoea-producing strains of *Esch.coli* were originally detected by their O (or somatic) antigens. Although the picture is not clear cut (i.e. there is overlapping) strains possessing certain O antigens tend to be found in each of the three main groups of diarrhoea-producing strains:
Enteropathogenic: 026, 055, 0111, 0119, 0127, 0128.
Enterotoxigenic: 06, 08, 025, 078, 0148, 0149
Enteroinvasive: 0124, 0136, 0144, 0164.

Note: most enterotoxigenic strains possess colonizing factor antigens: some strains of enteropathogenic *Esch.coli* elaborate enterotoxins.

Esch. coli Diarrhoea

Mainly seen as two types of disease:
1. *Infantile gastroenteritis*—largely confined to babies under 2 years of age.
2. *Travellers diarrhoea*—mostly in adults recently arrived in a foreign country.

1. Infantile gastroenteritis due to Esch. coli

Note: rota and other viruses are also an important cause of infantile gastro-enteritis.

Cause: generally the enteropathogenic strains listed above.

Clinical Features

Acute diarrhoea after an incubation period which varies from 1 to 3 days: the diarrhoea may lead to dehydration (Fig. 29.1) and acid base imbalance.

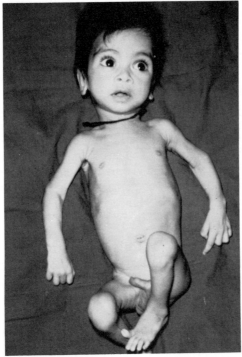

Fig. 29.1 Child with gastroenteritis. (Reproduced with permission from Abbott Laboratories 'Slide Atlas of Infectious Diseases' 1982, Gower Medical Publishing, London. Photograph by Professor H. Lambert).

Hypernatraemia is a particular problem because of the disproportionate loss of water relative to the sodium from the extracellular spaces.

Diagnosis

Isolation

Specimens: faeces

Culture: MacConkey's medium

Observe: and pick several pink (lactose-fermenting) colonies for further tests.

Identify: serologically with polyvalent then individual specific O antisera.

Treatment

Rehydration with correction of fluid loss, and electrolyte and acid-base imbalance: antibiotic therapy is of doubtful value—it may be useful in severe cases.

Epidemiology

Infection is sporadic in the community in Britain but the main problem in this country is outbreaks in nurseries.

Overseas in the Third World, in tropical countries with poor sanitation and housing, the disease is a major cause of infant mortality. Flies probably play a part in transmitting the infection.

Transmission: in nurseries and neonatal units, infection spreads directly from case to case and via environmental contamination (e.g. by communal toilets, shared equipment and the hands of the staff); in some neonates, their mother is the source of infection.

Control

Scrupulous hygiene in nurseries and neonatal units: examination of faeces of antenatal patients with diarrhoea and of all admissions to nurseries for presence of enteropathogenic serotypes.

Control of an outbreak: prompt isolation of cases and contacts; screening of staff to detect carriers of the epidemic strain; the unit should be closed to new admissions.

Overseas: control is only possible when clean water, sanitation and adequate housing can be provided.

2. Travellers diarrhoea (Turista)

Also known as 'Delhi belly', 'Montezuma's revenge', 'Tokyo two-step', etc.

Cause: generally the enterotoxigenic strains of *Esch.coli* described above. Most cases are not investigated bacteriologically.

Clinical features

Diarrhoea, abdominal pain and vomiting usually self-limiting and of a few days duration; occasionally protracted.

Route of infection: via contaminated food and drink; polluted water is the usual source of infection.

Diagnosis

Laboratory facilities are often not available at tourist resorts: information about the cause of this major problem to the tourist industries of many countries has come from a few research studies.

Control

Public health measures, e.g. clean water supply; caution in diet by travellers.

Other outbreaks of diarrhoea due to Esch.coli

Outbreaks in adults have been reported: for example, a large-scale outbreak of cheese-borne food poisoning has been attributed to *Esch.coli;* laboratory investigation of faeces for specific serotypes of *Esch.coli* is impractical in sporadic cases of diarrhoea.

STAPHYLOCOCCUS AUREUS

The classic example of toxic food-poisoning. Due to ingestion of food contaminated with the enterotoxin of *Staphylococcus aureus:* very rapid in onset because it is due to pre-formed toxin in food.

Cause: About 40 per cent of *Staph.aureus* strains produce one (or more) enterotoxins: there are five antigenically distinct enterotoxins — A, B, C, D, E: enterotoxin-producing strains mostly belong to phage group III.

Clinical features

Symptoms: acute onset of nausea and vomiting within a few hours of eating the contaminated food, sometimes followed by diarrhoea; self-limiting and rarely severe; dehydration is occasionally a problem in the uncommon severe case.

Pathogenesis

Pre-formed toxin: ingested in contaminated food has a local action on the gut mucosa. The toxin resists temperatures that kill *Staph. aureus* so that food may contain toxin but no viable staphylococci.

Diagnosis

Isolation

Specimens: the suspect food, vomit or faeces.

Culture: for *Staph. aureus* on ordinary media or mannitol salt agar (a good selective medium for the organism).

Identify: by coagulase test; later by phage typing to correlate identity of strains from food and patients.

Demonstration of enterotoxin

Specimens: food or vomit; culture-filtrate from an isolated strain of *Staph. aureus.*

Examine: by gel diffusion using specific antisera.

Note: Tests for toxin are only available at reference laboratories.

Treatment

The disease is short and self-limiting so treatment is unnecessary.

Epidemiology

Source of infection: is usually a staphylococcal lesion on the skin especially of the fingers of a food handler.

Route of infection: ingestion: enterotoxin-producing strains of *Staph. aureus* multiply in the food and liberate toxin: the toxin is relatively heat-stable and, unless the food is afterwards thoroughly heated, retains activity.

Food: usually cooked food. Contamination most often takes place after initial cooking: cold cooked meat is often implicated; unless food is correctly stored at 4°C, staphylococci can multiply in the warm conditions of the kitchen with consequent toxin production. Other foods often involved include dairy products, e.g. creams and custards.

Outbreak: the international nature of staphylococcal food-poisoning is vividly illustrated by a large outbreak in Japanese air travellers who became ill while en route from Tokyo to Denmark after eating ham contaminated by a food handler in Alaska.

Control

Food hygiene: exclusion of handlers with septic lesions; prompt refrigeration of food after preparation.

CLOSTRIDIUM PERFRINGENS

Diarrhoea or 'food poisoning' due to *Cl.perfringens* is fairly common and is due to contamination of food by spore-bearing (and therefore heat-resistant) anaerobic organisms.

Cause: Clostridium perfringens; classically, non-haemolytic strains which have particularly heat-resistant spores are involved but β-haemolytic strains with relatively heat-labile spores are being increasingly implicated.

Clinical features

Onset is acute—between 8 and 24 h after eating contaminated food: diarrhoea and abdominal pain are the predominant symptoms: vomiting is rare; the illness is self-limiting.

Pathogenesis

Heat-resistant spores of Cl.perfringens survive 100°C for 30 min: during cooling after cooking, spores germinate into vegetative bacilli. These multiply rapidly if food is stored at room temperature (the temperature range for growth of *Cl.perfringens* is 15–50°C) and if there are anaerobic conditions (e.g. some deep meat pies).

Following ingestion of food contaminated with vegetative *Cl. perfringens* sporulation takes place in the gut with liberation of a heat-labile enterotoxin which acts mainly on the small intestine.

Diagnosis

Strains which cause food-poisoning also form part of the normal flora in 5–30 per cent of people, so that laboratory diagnosis of an individual case is not possible. However, there are more than 60 serotypes of *Cl.perfringens* and isolation of the same serotype from most of the victims of an outbreak of food-poisoning and from suspect food (when available) is strong presumptive evidence that it is the cause.

Isolation

Specimens: faeces from as many as possible of the patients involved in the outbreak; samples of suspected food.

Culture: aminoglycoside blood agar anaerobically.

Observe: for typical colonies beta-haemolytic or non-haemolytic.

Identify: by Nagler reaction: serotype by slide agglutination in a specialist reference laboratory for epidemiology.

Treatment

Sometimes rehydration: antibiotic therapy is unnecessary.

Epidemiology

Sources of infection: Cl.perfringens is present in very large numbers as a commensal in the animal and human intestine; it is also ubiquitous in the environment.

Route of infection is the ingestion of food in which *Cl.perfringens* has multiplied: because it is an anaerobic organism, cooked and reheated meat pies, stews and gravies provide a suitable environment for bacterial multiplication.

Control

Food hygiene: especially adequate cooking of meat and meat products with prompt refrigeration if stored before consumption.

SECTION II: LESS COMMON BACTERIAL CAUSES OF DIARRHOEA

BACILLUS CEREUS

Diarrhoea due to this relatively non-pathogenic organism is typically associated with Chinese restaurants because of their frequent use of rice.

Cause: Bacillus cereus is an aerobic, spore-forming Gram-positive bacillus often found in soil, and the air and dust of the environment.

Clinical features

There are two distinct types of illness:
1. *Short incubation period:* 1–2 h, with nausea and vomiting often followed by diarrhoea; associated with bulk-prepared rice. This is by far the commoner type.
2. *Longer incubation period:* 6–16 h, sudden onset of abdominal pain and diarrhoea; associated with soups and sauces.

Pathogenesis

Unclear: possibly due to an enterotoxin.

Diagnosis

Isolation

   Specimens: suspected food, vomit, faeces,
   Culture: on ordinary media.
   Observe: for typical 'curled hair' colonies.

Treatment

The disease is self-limiting.

Epidemiology

   Source: B.cereus and its spores are widespread in soil and cereals are commonly contaminated with it.

   Route of infection: some spores survive cooking; if storage is at a warm temperature, there is germination into vegetative bacilli which multiply and produce toxin.

Control

   Food hygiene: correct storage of cooked food: reheating should be rapid.

CLOSTRIDIUM DIFFICILE

Antibiotic-associated colitis

Diarrhoea is a common side-effect of antibiotic therapy: usually mild and self-limiting but with occasional severe cases in which the colonic mucosa has the characteristic appearance of pseudo-membranous or antibiotic-associated colitis (Fig. 29.2).

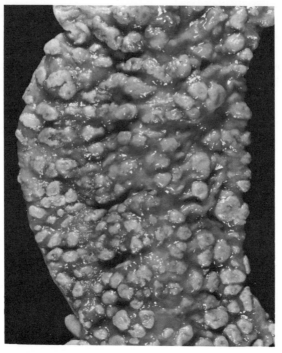

Fig. 29.2 Antibiotic-associated pseudomembranous enterocolitis. Segment of large bowel with typical pseudomembranous plaques. The plaques consist of fibrin and polymorphs and are the characteristic gross and microscopical lesions in the condition. (Photograph by Dr P.S. Macfarlane.)

Antibiotics: clindamycin, lincomycin and, to a lesser extent, ampicillin and tetracycline are most often implicated; many cases have followed administration of more than one antibiotic.

Cause: toxigenic strains of *Clostridium difficile*.

Clinical features

A life-threatening disease of the colon in which diarrhoea is the major

symptom: the diarrhoea varies but may be profuse (up to 20 stools per day): abdominal discomfort, fever and leucocytosis are often present.
Surgery: many cases follow abdominal operations.

Pathogenesis

Antibiotic therapy alters the balance of the normal flora and the changed environment may favour colonization, multiplication and toxin production within the colon by *Cl.difficile; Cl.difficile* is probably not part of the normal gut flora.

Diagnosis

Clinically, a high index of suspicion is necessary.

Proctosigmoidoscopy may show the characteristic membrane or isolated areas of white or yellow material adhering to the colonic mucosa: these areas should be biopsied for histological examination: the rectum is sometimes unaffected.

Laboratory diagnosis

1. *Direct demonstration of toxin* in faeces by testing for cytotoxic effect on cells in tissue culture, e.g. Vero or human embryo lung cells: *Faecal filtrate* is prepared and tested for cytotoxicity and for neutralization of the toxic effect by specific antiserum.
2. Isolation of *Cl.difficile* from faeces with subsequent demonstration of toxigenicity.

Treatment

Current antibiotic therapy must be stopped: oral vancomycin which is active against clostridia gives good results.

Control

Antibiotics should be limited to cases in which there is a clear indication for their use: clindamycin and lincomycin should be avoided if at all possible.

Neonatal necrotizing enterocolitis

This condition, which is not antibiotic-associated is an important and life-threatening complication in pre-term babies. No bacterial cause has been established.

YERSINIA

Yersinia are not a common cause of diarrhoea in Britain but have been reported with some frequency in Scandinavia and USA.

Causal organisms: Yersinia enterocolitica, Y.pseudotuberculosis.

Clinical features

A septicaemic illness with abdominal pain, diarrhoea and fever: diarrhoea is more prominent in infections due to Y.enterocolitica: yersiniosis may mimic appendicitis and the disease in fact is associated with acute *terminal ileitis* and *mesenteric lymphadenitis*.

Diagnosis

Isolation: from faeces or from blood cultures or mesenteric lymph nodes (if specimens of these are available).

Treatment

Tetracycline or aminoglycosides.

Epidemiology

Reservoir of infection: birds, wild and domestic animals.
Spread: via the faecal-oral route directly person-to-person or from animals; by contaminated food and water.

VIBRIO PARAHAEMOLYTICUS

This organism is a rare cause of infective diarrhoea transmitted by shellfish.

Cause: Vibrio parahaemolyticus—a marine bacterium found in warm coastal waters and in fish and shellfish.

Clinical features

Symptoms: acute onset of vomiting and diarrhoea usually 8-24 h after eating raw sea food—especially imported shellfish.

Epidemiology

More prevalent in warmer months. Most common in Far East where raw fish is a delicacy: cases in Britain have been due to imported sea food but the organism exists in British sea waters.

SECTION III: CHOLERA

In 1961, the world experienced the start of the seventh pandemic of cholera which has persisted and spread over the succeeding 22 years: the six previous pandemics were during the 19th and early 20th centuries. Glasgow Royal Infirmary is built on the site of mass cholera graves from the epidemic of 1849. The factors which determine the onset of epidemic spread are unknown.

Cause: Vibrio cholerae, two biotypes are recognised.
1. *Vibrio cholerae*—the cause of classical cholera and now replacing the El Tor strain as the predominant cause of cholera in Bangladesh.
2. *Vibrio cholerae El Tor*—responsible for the recent pandemic: causes a generally milder disease.

Clinical features

Acute onset—after an incubation period of from 6 h to 5 days—of abdominal pain and diarrhoea: the diarrhoea which is typically of exceptional severity, progresses to the continuous passage of *'rice-water' stools;* vomiting, dehydration, acidosis and collapse may follow; some cases, however, are much less severe with only mild diarrhoea.

Pathogenesis

Exotoxin: V.cholerae produces a potent protein exotoxin—very similar to the LT enterotoxin of *Esch. coli* enterotoxigenic strains—which is plasmid-coded: the toxin stimulates the activity of the enzyme—adenylcyclase—which raises the concentration of cyclic AMP in cells causing an increase in the flow of water and electrolytes into the bowel lumen; the fluid lost has relatively high concentrations of bicarbonate and potassium; *V.cholerae* is not invasive and does not penetrate the gut mucous membrane although adhesion to gut epithelium probably plays a part in its pathogenicity.

Diagnosis

Isolation

Specimen: faeces.
Culture: on selective medium (e.g. TCBS agar).
Observe: for typical colonies.
Identify: by slide agglutination with polyvalent antiserum.

Treatment

Correction of dehydration by the intravenous administration of fluid
and electrolytes to restore the acid-base balance: mortality can be
reduced from more than 50 per cent to nil with fluid replacement
treatment.

Tetracycline given orally or intravenously may help to limit the
duration of diarrhoea and reduce fluid loss.

Epidemiology

Reservoir of infection is the human gut.
Spread is faecal-oral usually via sewage contamination of the water
supply but sometimes via food contaminated by flies or unclean hands.
Symptomless carriers are common in epidemics—for example, the
case: carrier ratio with *V.cholerae El Tor* may reach 1:100: they form
the main source of infection.
Pandemic: since 1961, cholera El Tor gradually spread through the
Far East to reach Africa and beyond: air travel has probably increased
the risk of importation of the disease into cholera-free areas.

Control

Good sanitation with a clean water supply and adequate sewage
disposal together with personal hygiene (i.e. hand washing after
defaecation) are effective methods of controlling the spread of cholera:
unfortunately in many areas of the world these methods are simply not
practicable; carriers and cases should also, if possible, be isolated.

Vaccination

A vaccine containing heat-inactivated organisms is available: the
protection conferred is limited (as is the immunity following natural
infection).

Table 29.2 Non-bacterial causes of diarrhoea

Agent	Disease	Diagnosis	Treatment
Viruses			
Rotaviruses	Infantile gastroenteritis	Electron microscopy, ELISA	Symptomatic
Caliciviruses	Winter vomiting disease	Electron microscopy	Symptomatic
Astroviruses	Gastroenteritis	Electron microscopy	Symptomatic
Norwalk agent	Gastroenteritis	Electron microscopy	Symptomatic
Small round viruses	Gastroenteritis	Electron microscopy	Symptomatic
Protozoa			
Entamoeba histolytica	Amoebic dysentery	1. Microscopy of stools for trophozoites or cysts 2. Serology	Metronidazole
Giardia lamblia	Giardiasis	Microscopy of duodenal aspirate, stools for giardia trophozoites or cysts	Metronidazole
Cryptosporidium	Diarrhoea	Microscopy of stools for oocysts	Symptomatic

NON-BACTERIAL INFECTIVE DIARRHOEA

It must never be forgotten that diarrhoea is often due to infection with non-bacterial agents. The more important of these are listed with some of their characteristics in Table 29.2.

30

Botulism

━━━━━━━━━━━━━━━━━━━━━━━━━━━━━━━━━━━━━━━

A rare, severe disease due to ingestion of a bacterial toxin pre-formed in food: diarrhoea is not a symptom.
Causal organism: Clostridium botulinum, types A, B and E.

CLINICAL FEATURES

Incubation period: usually 12 to 36 h.

Symptoms: neurological—the toxin acts by reducing acetylcholine release at neuromuscular junctions resulting in neurological signs and symptoms such as oculomotor, pharyngeal paralysis, vomiting, constipation, thirst, dryness of mouth, vertigo; sometimes difficulty in speaking.

Prognosis: often fatal; death is due to respiratory failure.

PATHOGENESIS

Cl. botulinum is a spore-forming anaerobe which forms an exceedingly potent exotoxin and is found in soil, water, sludge; if spores contaminate food in conditions of anaerobiasis, germination follows and the vegetative bacilli multiply to produce toxin. Spore germination and toxin formation are inhibited by a low pH, e.g. they do not take place in acid fruits but can proceed in alkaline vegetables (e.g. string beans). The toxin is sensitive to heat and is destroyed by cooking.

DIAGNOSIS

1. *Specimen:* suspected food, patient's serum.
   Demonstrate toxin: by inoculation of mice.
   Observe: for paralysis and death.
   Identify: include in the test mice protected by antitoxins to types A, B and E toxins; protection by the appropriate antitoxin identifies the toxin present.

2. *Specimen:* suspected food.
   Isolate: Cl. botulinum by pasteurisation of the food to destroy non-sporing bacteria followed by anaerobic culture at 35 °C.
   Identify: suspect colonies of *Cl. botulinum.*

TREATMENT

Antibiotics are of no value: treatment is supportive with artificial ventilation, etc.; antitoxin may help to neutralise absorbed toxin.

EPIDEMIOLOGY

Table 30.1 shows the habitat, geography and the kind of food associated with the three types of *Cl. botulinum* that cause human disease.

Table 30.1 Medically important types of *Cl. botulinum*

Type	Habitat	Geography	Usual source of infection
A	Soil	USA USSR	Home-preserved vegetables, meat, fish
B	Soil	Europe USA	Meat, especially pork
E	Soil, sea water, sludge	Japan Canada Alaska	Raw or tinned fish

The disease is extremely rare in Britain but there was a small outbreak in Birmingham in 1978 due to tinned Alaskan salmon contaminated with type E toxin.

CONTROL

Amateur canning and preservation of food should be avoided. Commercial canning and pickling processes should be carefully controlled: a temperature that will kill the heat-resistant spores of *Cl. botulinum* is essential—e.g. 120 °C for 20 min.

31

Enteric fever

Enteric fever includes typhoid fever and paratyphoid fever. Both are due to salmonellae which are markedly more invasive than the salmonellae which cause food-poisoning. Paratyphoid fever is generally a milder disease than typhoid fever.

TYPHOID FEVER

Causal organism: Salmonella typhi.

CLINICAL FEATURES

Incubation period: 14-21 days; sometimes longer.

Symptoms: a septicaemic, febrile illness with headache, toxaemia, dullness, apathy. Rose-coloured spots (which contain the infecting organism) are often seen as a sparse rash on the abdomen. Splenomegaly is sometimes present. Generally some soft abdominal swelling and discomfort; leucopenia is common. Diarrhoea is a late symptom usually in the third week of illness.

Duration: untreated, about 4 weeks: symptoms clear up in around 3-4 days with antibiotic therapy.

Complications: relapse, intestinal perforation and haemorrhage are the most serious; rarely, periostitis, myocarditis, pneumonia.

Relapse: is common, with recrudescence of symptoms about a week after the end of the primary illness.

Carriers. Approximately 3 per cent of patients with typhoid fever become chronic carriers due to persistent infection of the gall bladder; this results in *S. typhi* being discharged into the gut and so into the faeces. Carriage is often associated with cholecystitis and gall stones.

PATHOGENESIS

The way in which *S. typhi* invades the body to cause disease is complex and is outlined diagrammatically overleaf:

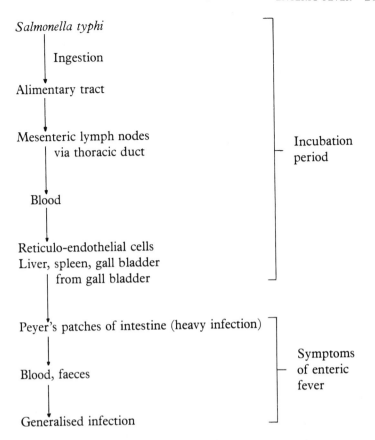

Salmonella typhi

| Ingestion

Alimentary tract

Mesenteric lymph nodes
    via thoracic duct   ─ Incubation period

Blood

Reticulo-endothelial cells
Liver, spleen, gall bladder
    from gall bladder

Peyer's patches of intestine (heavy infection)

Blood, faeces   ─ Symptoms of enteric fever

Generalised infection

PARATYPHOID FEVER

Causal organism: S. *paratyphi A, B and C; S. paratyphi B* is common in Britain; S. *paratyphi A and C* are virtually confined to tropical countries.

Clinically a milder febrile illness of shorter duration and incubation period; transient diarrhoea and symptomless infections are common. S. *paratyphi B* usually causes symptoms of diarrhoea rather than the classical syndrome of enteric fever.

Carriers: the percentage of cases who become carriers is lower than in typhoid fever.

Note: The text of the rest of this chapter applies to both typhoid and paratyphoid fevers.

DIAGNOSIS

Isolation

Specimens: faeces, blood, and urine. Blood culture is positive in over 80 per cent of patients in the first week of illness.

Faeces culture:

1. Solid selective media:
   a. MacConkey's medium
      Observe: for non-lactose-fermenting (pale) colonies
   b. Desoxycholate citrate agar
      Observe: for non-lactose-fermenting (pale) colonies
   c. Wilson and Blair bismuth sulphite agar
      Observe: for shiny black metallic colonies
2. Fluid enrichment media:
   a. Tetrathionate broth
   b. Selenite F
      Subculture: after 24 h onto MacConkey's medium
      Observe: for non-lactose-fermenting pale colonies.

Identification:

1. *Biochemical reactions* (API test)
   Note S. typhi, unlike other salmonellae (including *S. paratyphi*) produces no gas on fermentation of sugars.
2. *Serological:* preliminary identification with salmonella polyvalent 'H' and 'O' antisera. Final identification: send to Reference Typing Laboratory.

Bacteriophage typing: to distinguish between strains of *S. typhi* (and also of *S. paratyphi B*); useful for epidemiological investigation into the source of outbreaks.

Blood culture and urine culture: subculture blood cultures and culture urine directly on MacConkey's medium with subsequent identification as described above.

Serology

The classical Widal test is difficult to interpret especially if the patient has been immunised with typhoid vaccine. Diagnosis is best made by isolation of the infecting organism.

Widal test: agglutination test for antibodies to flagellar H antigens (using formalized bacteria) and somatic O antigens (using boiled bacteria) of *S. typhi, S. paratyphi A* and *B*.

Significant titre:

Titre: a titre of 80 to both H and O antigens suggests active infection in an unvaccinated patient living in an area where enteric fevers are not endemic.

Vaccinated patients: a considerable proportion of the population has been vaccinated against typhoid and paratyphoid fever: in them, diagnosis of infection by Widal test is unreliable. H antibodies persist longer than O antibodies.

Vi antibody: not normally tested although antibody to the Vi somatic antigen is usually produced during typhoid fever: Vi antigen can mask the O antigens so that Vi antibody is produced without production of O antibodies.

TREATMENT

Acute disease

1. *Chloramphenicol:* effective in clearing up symptoms rapidly; does not prevent relapse or carriage.
2. *Cotrimoxazole;* gave promising results in early trials; probably as good as chloramphenicol for rapid control of symptoms in acute phase.
3. *Ampicillin:* good in vitro activity is not matched by in vivo therapeutic effect.

Carriers

It is notoriously difficult to eradicate *S. typhi* from the gall bladder. Antibiotic therapy is effective in curing some carriers but, in a proportion, the infection persists and they become long-term and sometimes permanent, carriers.

1. *Ampicillin:* probably the most successful drug; treatment must be in large doses and prolonged.
2. *Chloramphenicol:* less successful in carriers than ampicillin; long courses of treatment in any event should be avoided because of risk of erythropoietic damage.
3. *Cholecystectomy:* cures carriage but the operation carries a small mortality rate and is only resorted to under special circumstances.

EPIDEMIOLOGY

Habitat: the human gut.

Source of infection. Chronic carriers who excrete the organism; excretion in faeces—less commonly in the urine—continues for about two months after the acute illness in most patients: after both typhoid and paratyphoid fevers a proportion of patients become permanent carriers.

Route of infection. Ingestion of water or food contaminated by sewage or via the hands of a carrier; direct case-to-case spread is very rare.

Infecting dose. Small numbers of *S. typhi* can cause typhoid fever— hence why water-borne infection is common despite the great dilution of organisms in sewage-contaminated water; large doses are required to infect in paratyphoid fever.

Sporadic cases are not uncommon in Britain: about three-quarters have acquired their infection abroad.

Outbreaks. Numerous—usually explosive—outbreaks are on record, several involving very large numbers of people. There are two main types.

1. *Water-borne:* in which sewage containing organisms from a carrier, pollutes drinking water, e.g. the outbreaks in Croydon 1937 and in Zermatt in 1963.
2. *Food-borne:* in which food becomes contaminated via polluted water or via the hands of carriers. 'Typhoid Mary', possibly the most famous carrier, worked as a cook in the USA and caused numerous outbreaks there in the early years of this century. Typhoid and paratyphoid bacilli multiply readily in most types of food.

Tinned food may become contaminated during canning, e.g. the large outbreak in Aberdeen in 1964 was due to a tin of corned beef which had been cooled in sewage-contaminated water; bacteria entered the can through tiny holes in the metal casing.

Shellfish often grow in estuaries where the water may be polluted by sewage: if eaten uncooked, they may cause infection; in the past a common source of typhoid, but not paratyphoid, fever.

Milk or cream products contaminated through handling by carriers have caused many outbreaks of both typhoid and paratyphoid fever: artificial cream is particularly associated with *S. paratyphi B* infection.

Other foods such as meat products, dried or frozen eggs, dried coconut, have been responsible for infection as a result of contamination by handlers who were carriers.

Animals: S. paratyphi B, unlike *S. typhi,* occasionally infects cattle; this has caused some outbreaks but much less commonly than infection from human sources.

CONTROL

Public health. The most effective way of controlling typhoid and paratyphoid fevers is provision of a clean water supply, adequate arrangements for sewage disposal and supervision of food processing and handling.

Carriers. If refractory to treatment, they must not be employed in food preparation. When instructed in personal hygiene (i.e. washing hands after defaecation), carriers are rarely a danger to family and close contacts.

Vaccine gives substantial but not solid protection.

Contains: heat-killed bacilli: the former 'TAB' vaccine with *S. typhi S. paratyphi* A & B is now replaced by monovalent vaccine containing only *S. typhi.*

Administered subcutaneously, in two doses.

32

Urinary tract infections

Urinary tract infections remain a major clinical problem 40 years after the introduction of antibiotics: many consultations in general practice are because of urinary infections.

Urinary infection is defined as bacteriuria, i.e. the multiplication of bacteria in urine within the renal tract: a concentration of 100 000 organisms per ml is regarded as *significant bacteriuria*.

Pyuria is the presence of pus cells (polymorphs) in the urine: it usually—but not always—accompanies bacteriuria:

Infections of the urinary tract may involve:
1. **Bladder**
   Cystitis.
2. **Kidney**
   pelvis—*pyelitis*
   parenchyma—*pyelonephritis*
   Note. It is difficult to distinguish between these two conditions and is probably better to refer to both as pyelonephritis. Pyelonephritis may become a chronic infection.
3. **Urethra**
   Urethritis (see Chapter 39).

CLINICAL FEATURES

1. Dysuria, frequency, urgency, suprapubic pain, sometimes haematuria are the classical symptoms of cystitis.
2. Loin pain and tenderness, rigors and fever are signs of pyelonephritis.
3. Chronic pyelonephritis causes general ill health and malaise with nocturia.

Women. Urinary infection is predominantly a disease of women (sex ratio 10:1).

Structural abnormalities of the renal tract, e.g. congenital, neurogenic bladder, prostatic enlargement, cause *obstruction* and *stasis* in the tract and greatly increase the risk of infection.

Symptoms of infection, usually those of cystitis, are surprisingly common: only a minority of women complaining of frequency and dysuria, perhaps one in ten, consult their doctors and of those who do, only about half have infected urine. Of the remainder, some will subsequently develop significant bacteriuria but others never have bacteriuria and the cause of their symptoms which may be recurrent, is unclear: this condition is often referred to as the 'urethral syndrome'.

Symptomless urinary infection, or 'covert bacteriuria', is, conversely, also not uncommon: it can be detected in 5 per cent of adult women, 1 to 2 per cent of girls and 0.3 per cent of boys (in whom it is often associated with abnormality of the renal tract).

Screening to detect, symptomless or covert bacteriuria has been advocated in schoolchildren and pregnant women.

CAUSAL ORGANISMS

Escherichia coli: the cause of 60-90 per cent of urinary infections.

Certain serotypes of Esch. coli are particularly common in urinary infection (e.g. 02, 04, 06, 07, 018, 075, etc.): this is probably because they are often present in the colon rather than because of inherently high pathogenicity for the urinary tract. However, some bacterial strains are reputed to be more invasive than others: factors associated with virulence include:

1. The possession of K (capsular) antigens which inhibit phagocytosis and the bactericidal effect of complement.
2. The ability to adhere to uroepithelium due to specialised fimbriae.

Staphylococcus saprophyticus: an important cause of infection related to sexual activity in women under 25 years old: detected in 30 per cent of such infections—surprisingly—seldom isolated from faeces and the anogenital region of young women.

Proteus mirabilis: responsible for 10 per cent of infections.

Klebsiella species: often multiply-antibiotic resistant.

Streptococcus faecalis: often found accompanying infection with coliforms.

Fastidious Gram-positive bacteria (e.g. lactobacilli, streptococci, corynebacteria) which require 48–72 h incubation in the presence of 7 per cent CO_2 for isolation. Routine urine culture fails to grow these organisms but when they have been detected by appropriate methods, it is claimed that they are associated with pyuria and symptoms. Infection with them may be one cause of the urethral syndrome.

Pseudomonas aeruginosa ⎤ especially after catheterisation
Staphylococcus aureus ⎦ or instrumentation
Mycobacterium tuberculosis: renal tuberculosis is described in Chapter 38.
Acute uncomplicated urinary infection is usually due to one type of organism.
Chronic infection is often associated with more than one type of organism.

SOURCE, ROUTE AND FACTORS INFLUENCING URINARY INFECTION

Source: the reservoir of urinary pathogens is the flora of the colon.

Route of infection is ascending via the urethra from the perineum.

Neonates: haematogenous spread may play a role in renal infection in the newborn.

Female preponderance is probably due to the shortness of the female urethra: turbulence of urinary flow during micturition may result in bacteria entering the bladder.

Colonisation of the periurethral area with potential pathogens is said to be a necessary prerequisite for infection: this may be prevented by the bactericidal activity of urethral and vaginal secretions.

Sterility of urine is maintained by 'flushing' (i.e. from the frequent and complete emptying of the bladder and constant in-flow of newly formed urine) and local defence mechanisms in the bladder wall.

Sexual intercourse is correlated with infection in young women — possibly due to retrograde 'milking' of the urethra during coitus ('honeymoon cystitis').

Incompetence of the vesico-ureteric valve: due to congenital abnormality or inflammation of the bladder wall causes reflux of urine into the kidney pelvis during micturition; this may lead to pyelitis and pyelonephritis.

DIAGNOSIS

Specimen: a mid-stream specimen of urine (MSSU) collected to avoid contamination from perineum or vagina: in babies, use a strategically placed self-adhesive plastic bag but suprapubic needle aspiration of the full bladder may be necessary. A catheter specimen of urine (CSU) is excellent but catheterisation to obtain urine for laboratory examination cannot be justified because of the risk of introducing infection.

Transport: bacteria multiply in urine so specimens must be submitted to the laboratory within 2-4 h of collection; if this is not possible, do one of the following:

1. Refrigerate the specimen at 4 °C.
2. Use container with boric acid, a bacteriostatic preservative, to give a final concentration in urine of 1.8 per cent.
3. Use a dip slide coated on both sides with culture medium and inoculated by dipping into the urine: can be read even after several day's delay.

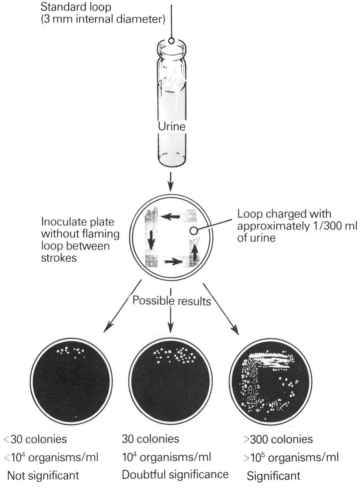

Standard loop
(3 mm internal diameter)

Urine

Inoculate plate
without flaming
loop between
strokes

Loop charged with
approximately 1/300 ml
of urine

Possible results

<30 colonies | 30 colonies | >300 colonies
<10⁴ organisms/ml | 10⁴ organisms/ml | >10⁵ organisms/ml
Not significant | Doubtful significance | Significant

Fig. 32.1 Diagram illustrating the semi-quantitative culture of urine specimens.

Direct examination

Wet film:	for the presence of pus cells and bacteria: erythrocytes and casts should be noted.
Gram film:	not often required: sometimes useful for preliminary identification of bacteria. Plate 13 shows a typical Gram film of the urinary deposit in an acute infection due to Gram-negative bacilli.

Culture: semi-quantitative culture on media such as CLED or Mac-Conkey agar (to prevent swarming of proteus).

Observe and count the number of colonies obtained (see Fig. 32.1 and 2).

1. *More than 100 000 bacteria per ml:* evidence of urinary infection; carry out sensitivity tests with appropriate antibiotics.
2. *Between 10 000 and 100 000 bacteria per ml:* significance doubtful; further specimens should be obtained.
3. *Less than 10 000 bacteria per ml:* regard as contaminants (unless the patient is on treatment for a known urinary infection)

Note. One cause of pyuria without bacteriuria is *renal tuberculosis:* when there is no obvious cause for this finding (e.g. recent urinary infection), three entire early morning specimens should be examined for *Mycobacterium tuberculosis.*

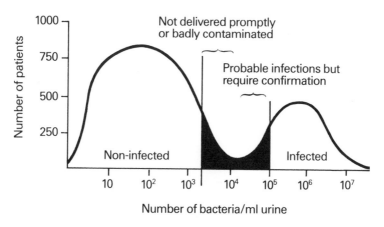

Fig. 32.2 Significance of urinary bacterial counts (Courtesy of Professor W. Brumfitt).

TREATMENT

Several suitable oral antibiotics are available: all are excreted in the urine in high concentration:

sulphonamide	nitrofurantoin
trimethoprim	nalidixic acid
cotrimoxazole	cephalexin or cephradine
ampicillin	

Chronic pyelonephritis: parenteral therapy with an aminoglycoside or cephalosporin may be required.

Cure rate of the acute infection is around 90 per cent but 50 per cent of the infections recur within a year: about 10 per cent of patients have repeated recurrences, most likely if infection involves the kidney.

Relapse: recurrence within one month of stopping treatment is usually due to relapse—i.e. recrudescence of infection with the same organism: evidence of therapy failure.

Reinfection: later recurrence is often a result of reinfection with a different organism.

Detection of site of infection within urinary tract: a number of laboratory tests, none foolproof, have been proposed to distinguish between renal and bladder bacteriuria. Kidney involvement is said to be likely when:

(i) renal concentration efficiency is reduced—tested by water deprivation

(ii) excess of $ß_2$ microglobulin is present in urine

(iii) antibodies to the infecting organism can be detected in serum

(iv) antibody-coated bacteria can be demonstrated in the urine by immunofluorescence microscopy.

Recurrent infections may indicate underlying abnormality of the urinary tract and require further investigation—e.g. urogram; management often requires long-term suppressive treatment with low doses of cotrimoxazole or nitrofurantoin.

In adults: bacteriuria—with or without symptoms— is nevertheless a relatively benign condition and permanent or progressive renal damage is rare.

In young children: under 5 years of age, the prognosis is much worse: bacteriuria with vesico-ureteric reflux often results in progressive renal damage with scar formation and impairment of kidney growth; some of these children go on to develop *chronic pyelonephritis*—which accounts for 20 per cent of end-stage renal failure.

SURGICAL URINARY INFECTION

A particular problem in urological and gynaecological wards usually following catherization and instrumentation.

Catheterization: the risk of infection after a single catheterization— even if carefully carried out—is 5 per cent: almost all patients with an indwelling catheter develop infection.

Source of infection

1. *Endogenous:* from contamination of the patient's urethra or perineum by bacteria from the colonic flora.
2. *Exogenous:* due to cross-infection with bacteria from the infected urinary tract of another patient: transmission by instruments (e:g. cystoscopes, catheters) or hands of doctors and nurses.

Prevention

Careful attention to aseptic technique: use of disposable plastic catheters; introduction of antiseptics into the urethra before instrumentation.

Indwelling catheters should be attached to a closed drainage system to prevent retrograde bacterial spread into the bladder from the collection bag.

Prophylactic antibiotic therapy may be given to cover the immediate post-operative period and reduce the risk of septicaemia developing from an infected urinary tract.

33

Meningitis

Bacterial meningitis (classically 'pyogenic' or polymorphonuclear) is a much more severe disease than viral meningitis (classically 'aseptic' or lymphocytic): despite antibiotic therapy bacterial meningitis remains a serious cause of morbidity and mortality (Fig. 33.1) and is a bacteriological emergency requiring prompt diagnosis and treatment.

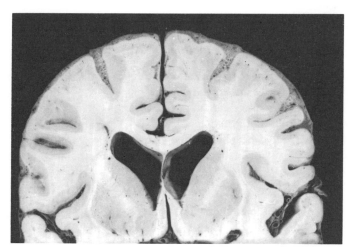

Fig. 33.1 Bacterial meningitis. Section of brain showing a thick layer of purulent exudate surrounding the brain and extending into the sulci. (Photograph by Professor J. Hume Adams.)

CLINICAL FEATURES

Symptoms: severe headache with malaise and fever; the onset is often abrupt: vomiting, photophobia and convulsions are not uncommon and patients may show irritability, apathy or drowsiness progressing to unconsciousness.

Signs of meningeal irritation—neck and spinal stiffness: pain and resistance on extending the knee when the thigh is flexed (Kernig's sign).

'*Meningism*' (apparent signs of meningeal irritation without meningitis) may be a feature of other types of severe infection: the differential diagnosis of meningitis has to be made from this and from subarachnoid haemorrhage and cerebral abscess.

Neonatal meningitis: the characteristic clinical features of meningitis are usually absent—the only presenting features being that the baby is obviously unwell with failure to feed and often vomiting. The condition is eight times more common in premature than in full-term babies.

Meningitis in the elderly and the immunocompromised: typical clinical signs and symptoms may again be absent; mental confusion may be a prominent feature.

Sequelae

Unfortunately, despite the advent of antibiotic therapy, neurological sequelae are not uncommon in survivors: the main types of sequelae are:
1. Encephalopathy—a state of altered cerebral function (e.g. mental retardation, drowsiness) and convulsions.
2. Cranial nerve palsies.
3. Cerebral infarction or abscess.
4. Obstructive hydrocephalus either external due to blockage of CSF drainage through arachnoid granulations or internal involving the ventricles.
5. Subdural effusion of sterile or infected fluid.

Age: Although meningitis is largely a disease of infancy and childhood, the disease is encountered throughout life.

CAUSAL ORGANISMS

1. Neisseria meningitidis (meningococcus)

The main cause of meningitis in Britain: affects all ages but most common in children and young adults. Of the eight serogroups of meningococci, group B, followed by Groups A and C, strains are responsible for most disease, the other groups being commoner in carriers than cases. Meningococcal meningitis is endemic and sometimes epidemic.

Incubation period: short, around 3 days.

Source: the reservoir is the human nasopharynx.

Spread: via infected respiratory secretions (i.e. 'droplet spread') from carriers and cases. Carriage rate in normal populations is about 5 to 10 per cent: this may rise to 50 per cent or more in household contacts of sporadic cases and during epidemics in closed communities, e.g. institutions, military camps.

Route of infection: from nasopharynx via blood stream to meninges. During the *meningococcaemia* a characteristic *petechial rash,* rare in other types of meningitis, is a common finding. Possession of specific antibody prevents the meningococcaemia and meningitis.

Control: avoid overcrowding in living and working conditions.

2. Haemophilus influenzae

Commonest cause of meningitis in infants and pre-school children aged 1 month to 4 years. The causal strains are capsulated and serologically almost always of Pittman type b. A nasopharyngitis precedes spread to the blood stream followed by involvement of the meninges.

3. Streptococcus pneumoniae (pneumococcus)

Although not uncommon in children this is the usual cause of meningitis in the middle-aged and elderly, often in patients who are in poor general health; this type of meningitis is often the sequel to pneumococcal infection of the lungs, sinuses or middle ear.

4. Mycobacterium tuberculosis

Seen in people of all ages and can occur at any stage after the primary infection. Now rare in developed countries.

Mortality rate

The overall case fatality rate in meningococcal and haemophilus meningitis is around 5 per cent. Pneumococcal and tuberculous meningitis have a higher mortality of the order of 20 per cent.

Rare causal organisms:

1. *Listeria monocytogenes:* mainly seen in infants, the elderly and in immunocompromised patients.

2. *Cryptococcus neoformans:* meningitis due to this yeast is rarely seen in previous normal people but is found in immunocompromised patients, e.g. those with leukaemia or lymphoma.
3. *Leptospira interrogans.*

Neonatal meningitis: is mainly due to Gram-negative bacilli such as *Escherichia coli, Klebsiella species* and *Proteus species:* ß-haemolytic streptococci of Lancefield group B are another important cause and are usually acquired from the mother's vagina. The baby is at greatest risk if there has been prem ure rupture of the membranes during labour or if the mother has a perinatal genital tract infection.

PATHOGENESIS

Spread to the CNS is by one of two routes:
1. *Haematogenous* via blood stream to the subarachnoid space.
2. *Direct* through the skull from infected foci in such sites as the middle ear, paranasal sinuses and nose via the cribriform plate.
   Controversy exists as to which is the more important route: probably haematogenous but there may be direct spread when there is an adjacent focus of infection or following skull fracture.

DIAGNOSIS

Laboratory diagnosis of bacterial meningitis depends on examination of the CSF. Table 33.1 lists the results of CSF examination in meningitis compared to that in the two most important diseases in the differential diagnosis i.e. subarachnoid haemorrhage and cerebral abscess.

Red blood cells may be present in a normal CSF due to accidental damage to a blood vessel during lumbar puncture (universally known as a 'bloody tap')—in such cases the supernatant fluid after centrifugation is clear whereas in subarachnoid haemorrhage it is stained yellow to orange (xanthochromia).

Isolation

1. *Specimen:* CSF obtained by lumbar puncture.
Examine:
a. In a counting chamber for nucleated cells and erythrocytes.
b. Gram film of centrifuged deposit for bacteria and cells.
c. If indicated, Leishman film of centrifuged deposit to differentiate polymorphonuclear leucocytes from lymphocytes.

d. If indicated, Ziehl-Neelsen film of centrifuged deposit for tubercle bacilli.

Culture:

a. Centrifuged deposit onto blood-agar, chocolate agar and into glucose broth and cooked meat broth: incubate plates in air plus 5 per cent CO_2.

b. If indicated, Lowentein-Jensen medium for culture for tubercle bacilli.

2. *Specimen:* blood culture; positive in over 40 per cent of patients with meningitis due to *N. meningitis, H. influenzae* and *Strep. pneumoniae.*

Demonstration of bacterial products:

Of value when meningitis has been partially treated and no infecting organisms can be seen or cultured.

a. *Bacterial antigen:* detect by an immunological method e.g. countercurrent immunoelectrophoresis, latex agglutination, coagglutination using specific antisera to the common serogroups of *N. meningitidis, H. influenzae* type b and a pooled antipneumococcal serum.

b. *Bacterial endotoxin:* detect by the Limulus test (p. 250): may be positive in meningitis caused by Gram-negative bacteria.

ANTIBIOTIC TREATMENT

Initial treatment should be parenteral by the intramuscular or intravenous route: later on antimicrobial drugs can be administered orally. Opinions differ on the need for intrathecal drug administration.

1. *Meningococcal meningitis*

Penicillin is the drug of choice: sulphonamide penetrates the CSF well and was formerly widely used but its effectiveness has diminished due to the increased frequency of sulphonamide-resistant strains—in the UK at least 10 per cent of strains are fully resistant and a further 20 to 60 per cent are partially resistant.

Chemoprophylaxis is indicated for close contacts of a patient to eradicate the organism from the nasopharynx. Use sulphonamide (if the strain is sensitive) or rifampicin or minocycline (a tetracycline). Note that pencillin is *not* effective in this situation.

Immunoprophylaxis: immunogenic polysaccharides to a number of serogroups—A, C, W135 and Y—are available but unfortunately it

Table 33.1 Findings in cerebrospinal fluid

	Causal microorganisms	Appearance	Cells per mm³	Microbiology	Protein	Glucose
Normal	—	Clear, colourless	0-5 lymphocytes	Sterile	150—450 mg per litre	2.8—3.9 mmol per litre
Bacterial meningitis	*Neisseria meningitidis, Haemophilus influenzae, Streptococcus pneumoniae*	Turbid	500–20 000 mainly polymorphs, few lymphocytes	Bacteria in Gram-stained deposit. Growth on culture	Markedly raised	Reduced or absent
Viral (aseptic) meningitis	Enteroviruses, Mumps virus	Clear or slightly turbid	10–500 mainly lymphocytes	Viruses rarely isolated from CSF Diagnosis by stool culture (enteroviruses) or serology (mumps)	Normal or slightly raised	Normal
Tuberculous meningitis	*Myobacterium tuberculosis*	Clear or slightly turbid	10-500 mainly lymphocytes, polymorphs in early stages	AAFB in Z.N. stained deposit – often scanty. Growth on L.J. culture	Moderately raised	Usually reduced
Cerebral abscess	*Streptococcus milleri, Bacteroides species, Staphylococcus aureus, Proteus species*	Clear or slightly turbid	0–500 mainly polymorphs, some lymphocytes	Organisms often not present in CSF	Normal or raised	Normal
Subarachnoid haemorrhage	—	Turbid, often blood-stained. Supernatant yellow-orange	Large numbers of red blood cells	Sterile	Markedly raised	Normal

has not been possible to prepare a satisfactory vaccine to group B strains. Vaccines are not of value in the UK where the majority of infections are caused by group B strains and, in addition, are sporadic. Their use is in epidemics in other parts of the world which are often caused by group A and group C strains.

2. Haemophilus influenzae meningitis

Chloramphenicol penetrates readily to the CSF and is the drug of choice. Ampicillin is less effective and resistant strains of *H. influenzae* may be encountered nowadays. Do not give these drugs in combination.

3. Pneumococcal meningitis

Penicillin is most widely used; the alternative is chloramphenicol.

4. Tuberculous meningitis

Triple therapy with isoniazid, rifampicin and ethambutol; pyrazinamide may be added to this regimen initially because of its excellent penetration into CSF. Treatment should be continued for a year.

5. Neonatal meningitis

Most often due to coliform bacilli: give either: (i) gentamicin and ampicillin or (ii) chloramphenicol.

Group B β-haemolytic streptococcal meningitis in babies is best treated with penicillin and gentamicin in combination.

Gentamicin penetrates the blood-brain barrier poorly and may be given intrathecally or intraventricularly as well as systemically.

6. Meningitis of unknown cause

When many polymorphs are present but no bacteria have been detected in the deposit of the CSF and there is no growth on culture, treatment must be empirical. This state of affairs is usually the result of inadequate treatment given outside hospital before the patient is admitted. Broth cultures may give a positive result after a few days' incubation when cultures on solid media remain negative.

On a *best-guess* basis give either chloramphenicol alone or triple therapy of penicillin, sulphonamide and chloramphenicol. Triple therapy, once very popular, is used less often nowadays.

34

Sepsis

An ill-defined term, *sepsis* covers numerous and diverse purulent infections, some trivial and others serious: they include superficial skin infections, wound infection, peritonitis and abscesses. These diseases are amongst the most common encountered in medicine and dealt with in bacteriology laboratories. The causes are also numerous and involve a large number of different bacteria.

SKIN INFECTION

The skin is an efficient barrier to infection and, provided that it is not breached, usually prevents invasion either by resident commensal or exogenous bacteria. Nevertheless skin infections are common probably because minor skin trauma is a part of everyday life.

The main forms of skin sepsis are shown in Table 34.1.

Table 34.1 Skin infections

Infection	Site	Causal organism
Boil	Hair follicle	*Staph. aureus*
Carbuncle	Multiple hair follicles	*Staph. aureus*
Stye	Eyelash follicle	*Staph. aureus*
Sycosis barbae	Shaving area	*Staph. aureus*
Impetigo	Cheeks, around mouth	*Staph. aureus* *Strep. pyogenes*
Erysipelas	Face, sometimes limbs	*Strep. pyogenes*
Pemphigus neonatorum	Infant's skin	*Staph. aureus*
Toxic epidermal necrolysis	Infant's skin	*Staph. aureus*
Acne vulgaris	Face and back	*P. acnes*

CLINICAL FEATURES

Boils, carbuncles, styes, sycosis barbae

With the exception of carbuncles, these skin infections are uncomfortable and unsightly rather than serious. Carbuncles are rarely seen nowadays except in diabetics who have a predisposition to develop septic lesions; they are associated with considerable malaise and constitutional disturbance.

Below are listed some characteristics of these forms of sepsis.

1. Due to *Staphylococcus aureus* mainly of phage groups I or II.
2. Tend to be recurrent—appearing in crops at the same site often over weeks or months.
3. Infection is usually endogenous due to a strain carried in the nose and on the skin.
4. Generally commoner in males than females: surprisingly often seen in previously healthy young males; sycosis barbae is a chronic infection of the skin of the shaving area and consists of a septic postular rash: almost certainly spread by the minor trauma made to skin and hair follicles by a razor: seen only in males (Fig. 34.1)

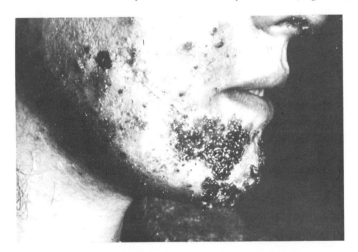

Fig. 34.1 Sycosis barbae. Staphylococcal infection of the skin of the shaving area. (Photograph by Dr A. Lyell).

Treatment: boils and styes do not require antibiotic therapy and, in any event, this does not prevent recurrences. Severe infections like carbuncles should be treated with penicillin if the causal strain is sensitive; otherwise with a different antistaphylococcal drug such as flucloxacillin. Sycosis barbae should be treated with topical antibiotics e.g. cream containing neomycin and bacitracin or fusidic acid.

Carriage sites: application to the nostrils of creams containing antiseptics, e.g. chlorhexidine or antibiotics such as neomycin or bacitracin suppresses (but rarely eradicates) nasal carriage of *Staph. aureus;* regular use of hexachlorophane soap reduces overall skin carriage.

Impetigo

A disease of young children in which vesicles appear on the skin around the mouth later to become purulent with characteristic honey-coloured crusts: nowadays most often due to *Staph. aureus* but *Streptococcus pyogenes* can also cause impetigo (Plate 18).

Outbreaks are not uncommon in schools where infection is spread by contact or via shared towels and other contaminated fomites.

Glomerulonephritis has been described following impetigo due to *Strep. pyogenes* especially Griffith type 49.

Treatment: antibiotics topically, e.g. tetracycline, chloramphenicol, bacitracin; systemic antibiotics are rarely required for this superficial infection. *Note.* Avoid the local application of penicillin which is liable to cause hypersensitivity.

Erysipelas

A spreading infection due to *Strep. pyogenes* presenting as a red, indurated and sharply demarcated area of skin: oedema often develops causing a characteristic 'orange skin' texture to the skin; the patient may be acutely ill with high fever and toxaemia. A rare disease nowadays.

Treatment: penicillin.

Neonatal skin sepsis

Neonates are very susceptible to infection and *Staph. aureus* can become epidemic in neonatal nurseries with outbreaks of pustules, sticky eyes, boils and abscesses. Neonates readily become carriers in nose, skin and umbilical stump.

Two severe infections associated with skin splitting are sometimes seen in neonates and infants; both are due to staphylococci of phage group II which produce an epidermolytic toxin which causes skin splitting and desquamation.

1. *Pemphigus neonatorum:* in which the skin splitting is focal or

localised in large vesicles or bullae; although a more serious disease than the neonatal infections listed above, it responds well to antibiotic treatment.

2. *Toxic epidermal necrolysis,* also called *Ritter-Lyell disease* or, more descriptively, 'scalded skin syndrome': a serious disease in which large areas of the skin desquamate leaving a red weeping surface which resembles a scald (Plate 19). Seen predominantly in neonates but also in young children and occasionally in adults: usually responds to antibiotic treatment with recovery.

Treatment: flucloxacillin.

Toxic shock syndrome

Seen mainly in menstruating women and associated with tampons which encourage multiplication of *Staph. aureus* present as a commensal in the vagina. Presents with fever, collapse and a skin rash with desquamation: there is usually multi-system involvement. Occasionally seen as a complication of other forms of staphylococcal sepsis.

Due to a toxin, toxic shock syndrome toxin-1 (TSST-1) elaborated locally but absorbed into the blood stream to produce widespread effects.

Phage group: most of the *Staph. aureus* strains responsible are in phage group I, occasionally phage group II.

DIAGNOSIS OF SKIN INFECTIONS

Specimens: swabs from lesions: pus, exudate.

Direct Gram film: observe for bacteria, noting especially the arrangement of any Gram-positive cocci (Plates 2 and 3).

Culture: blood agar, (i) aerobically and (ii) anaerobically—for enhanced growth with better demonstration of β haemolysis by *Strep. pyogenes.*

Observe: for typical colonies.

Identify: as appropriate for the organisms isolated.

Acne vulgaris

A common and disfiguring skin disease of adolescence: sometimes persists into adult life leaving residual pitting or scarring; probably not primarily an infectious disease but bacteria probably play a role in its pathogenesis.

Propionibacterium acnes (and *P. granulosum*) can regularly be isolated from inflamed comedos (whiteheads): these bacteria probably induce an inflammatory reaction in the skin by the production of lipase which liberates irritant fatty acids from lipid in sebum within sebaceous glands.

Treatment: long-term tetracycline in severe cases.

Note: Bacteriology plays no part in diagnosis.

CELLULITIS

An infection of subcutaneous tissue: there are two main clinical and bacteriological forms:

1. *Acute pyogenic cellulitis* due to *Strep. pyogenes:* presents as a red painful swelling usually of a limb and commonly associated with lymphangitis and lymphadenitis involving local draining lymph glands (Plate 20).
   Treatment: penicillin.

2. *Anaerobic cellulitis:* a rare infection sometimes due only to non-sporing anaerobes (e.g. bacteroides) or clostridia but more usually a synergistic infection with both aerobic and anaerobic bacteria; the aerobes produce reducing or anaerobic conditions which enable the anaerobes to multiply.

   Causal organisms: a combination of *aerobes* (coliforms, *Pseudomonas aeruginosa, Staph. aureus, Strep. pyogenes*) and *anaerobes* (most often bacteroides, anaerobic cocci—rarely clostridia).

   Clinically: there is redness, swelling and oedema, around the primary wound (which may be traumatic or surgical): usually situated on the abdomen, buttock or perineum, less commonly on the leg. Two syndromes may develop:

   a. *Progressive bacterial gangrene:* in which the skin rapidly becomes purple in colour and develops central necrosis (Plate 21). When this condition affects the penoscrotal region it has been called 'Fournier's gangrene'.
      Treatment: surgery and appropriate antibiotics.

   b. *Necrotising fasciitis (Meleney's gangrene):* in which the external appearance of the skin remains normal initially while the infective-ischaemic process spreads along the fascial plane causing extensive necrosis. Later the overlying skin, deprived of its blood supply, becomes painful and red and, finally, numb and necrotic. The patient is severely ill with fever and toxaemia, sometimes with septic shock.
      Treatment: wide excision of the skin to expose the entire area of

necrotising fasciitis and general supportive measures including appropriate antibiotics.

WOUND INFECTION

Minor wounds are commonplace in everyday life and, naturally, some become infected. *Staph. aureus* is the commonest cause: most heal without antibiotic treatment.

HOSPITAL WOUND INFECTION

Clinically, a much more serious problem. Obviously the majority of *surgical patients* have wounds which involve not only skin and subcutaneous tissue but also muscle and deeper tissues including sometimes bone, peritoneum and viscera.

CLINICAL FEATURES

Surgical wound infection

Presents as reddening of wound edges usually with pus formation: sometimes pus gathers below the suture line in the deeper layers of the wound to form a *wound abscess;* as a rule this eventually discharges to the surface through the sutured incision.

Patients may develop fever but constitutional disturbance is often minimal: the main problem is that infection impairs and delays healing and therefore prolongs hospitalisation.

Complications:

1. *Dehiscence:* the wound may break down completely—sometimes with exposure of viscera, etc.—and require to be resutured.
2. *Spread of infection* to:
   a. *Local tissues*—e.g. the peritoneum in the case of abdominal wounds.
   b. *Blood*—causing septicaemia.

The main bacterial causes of surgical wound infection are shown in Table 34.2

Sources:

Endogenous infection: a considerable proportion of wound infections are due to organisms carried in the patient's skin e.g. *Staph. aureus* or

Table 34.2 Main bacterial causes of surgical wound infection

Bacteria	Species	Most common site
1. Aerobic	*Staph. aureus*	Any wound
	Esch. coli	Any wound but
	Proteus species	especially abdominal,
	Klebsiella species	urological and
	Strep. faecalis	gynaecological
	Ps. aeruginosa	Urological; burns
2. Anaerobic	*B. fragilis*	Abdominal and
	B. melaninogenicus	gynaecological
	Anaerobic cocci	
	Cl. perfringens	

bacteria such as coliforms or bacteroides present in the gut as commensals. During hospitalisation before surgery the patient may become colonised with antibiotic-resistant bacteria from the environment and these may later be responsible for wound infection: this represents endogenous infection from a hospital source.

Exogenous infection of wounds can be acquired during operation from the surgeon or other theatre personnel.

Pseudomonas aeruginosa is a Gram-negative bacillus which is rarely found in the human bowel: often highly antibiotic-resistant, it is usually acquired exogenously from contamination of the environment e.g. water, fluids, ventilators, humidifiers—even antiseptic solutions.

Mixed infection: surgical wounds are often infected with more than one bacterial species (except in the case of *Staph. aureus* which is usually present in pure culture): for example, it is common to find more than one species of coliform together with *Strep. faecalis* and *Bacteroides species* or *Clostridium perfringens* this flora is known to bacteriologists as *faecal flora* and is the usual cause of abdominal wound sepsis. Note that the presence of *Cl. perfringens* does not imply that gas gangrene is about to supervene. The pathogenic potential of *Bacteroides species* and anaerobic cocci was underestimated in the past: their establishment in a wound may be aided by the presence of coliforms which use oxygen to promote anaerobic conditions and also by blood and damaged tissue.

Factors affecting surgical wound sepsis

The following factors increase the likelihood of wound infection:
1. Operations involving opening of the bowel.
2. Presence of drains or other foreign bodies.
3. Long operations.

4. Large wounds with considerable tissue trauma.
5. Obesity.

Incidence: The sepsis rate of surgical wounds therefore varies:
Low: in clean elective surgery e.g. hernia repairs, 'cold' orthopaedic operations—around **1 per cent.**
High: in operations and particularly emergency surgery in a contaminated site (e.g. large bowel)—around **15 per cent.**

Prevention

Difficult and complex: rigid observance of aseptic and antiseptic technique both in preparation for and during operation; many patients are colonised before operation with organisms which subsequently infect their wounds—the risk of this is increased by length of stay in hospital.

Theatre: surgeons and assistants must wear gowns to prevent bacterial dispersal via squames shed from the body surface and masks to trap respiratory secretions; surgeon's hands and patient's skin must be disinfected before operation; the surgeon and his assistants must wear gloves; theatres should be kept free from dust and have positive pressure ventilation to prevent air and dust being sucked in from outside; they should be sited away from main thoroughfares in the hospital.

Infection acquired at operation is most often due to bacteria entering the wound either from the patient's commensal flora or from the skin of the surgeons and assistants.

Ward: bacteria abound in hospital wards: patients with discharging wounds should be isolated to prevent the dissemination of pathogenic bacteria; floors must be kept clean and free from dust; blankets, now made from cellular cotton and not wool, must be regularly washed at a sufficiently high temperature to kill vegetative bacteria.

Antibiotic prophylaxis: if carefully chosen, antibiotics given peroperatively and for a short time thereafter can reduce the incidence of wound infection after certain operations; especially indicated when heavy soiling of the wound edges is unavoidable, e.g. in colonic surgery.

SPECIAL TYPES OF WOUND INFECTION

1. Burns

Extensive burns consist of a large moist exposed surface which is usually being prepared for a skin graft; always heavily colonised with bacteria—surprisingly this often does not prevent the graft taking.

Two organisms present particular problems in burns units:-

(i) *Strep. pyogenes,* a highly dangerous organism in a burns unit, not only affects the patient's general well-being, e.g. by causing septicaemia, but also causes the graft to fail.

(ii) *Pseudomonas aeruginosa* has a particular propensity to persist in a burns unit: of relatively low invasive powers, it can be difficult to eradicate both from an individual patient and from the environment of the unit.

Septicaemia: the invasion of the blood stream with bacteria from the burn is a dangerous complication of infected burns.

2. Orthopaedics

Wound infection is also a problem in orthopaedic surgery where healing of bone by first intention is important for future weight-bearing and movement. *Staph. aureus* is the principal cause.

Hip replacement has revolutionised the mobility of elderly patients with arthritis: about 0.5 to 2 per cent of these operations fail due to low-grade chronic infection which causes the prosthesis to work loose; most often due to *Staph. epidermidis.*

3. Puerperal sepsis (including septic abortion)

Formerly due mainly to *Strep. pyogenes* the important causes nowadays are streptococci of Lancefield group B, bacteroides (especially *B. melaninogenicus*) and anaerobic cocci; still a severe disease which requires prompt treatment.

Cl. perfringens is a rare but dangerous cause of puerperal sepsis.

4. Tracheostomies

Relatively common in modern hospitals—especially with the introduction of intensive-care units; the wounds have a marked tendency to become colonised with coliforms—which are often multiple-antibiotic resistant—from the hospital resident flora. Low-grade pathogens like *Acinetobacter* and *Serratia species* can be particularly difficult to eradicate; it is often uncertain if the presence of these and other coliforms in tracheal secretions actually harms the patient.

DIAGNOSIS OF WOUND INFECTION

Specimen: swab of pus, exudate, tissue from wound.

Direct Gram film: observe for organisms present.

Culture: blood agar, CLED or MacConkey agar aerobically; blood and aminoglycoside blood agar anaerobically; Robertson's meat medium.

Observe and identify: the bacteria isolated by standard tests.

TREATMENT

Antibiotics appropriate for the bacteria isolated but may not be necessary in the absence of generalized symptoms.

PERITONITIS

Another and serious form of sepsis: often a complication of abdominal surgery and of diseases that cause perforation of the gastrointestinal tract, e.g. peptic ulcer, diverticulitis, acute appendicitis, Crohn's disease: perforated typhoid ulcer is still a cause of death in enteric fever.

CLINICAL FEATURES

The patient's condition deteriorates markedly with fever, toxaemia and shock: there may be tenderness on palpation of the abdomen and absence of bowel sounds due to paralytic ileus.

Causal organisms: usually a mixed infection with 'faecal flora', i.e. coliforms, *Strep. faecalis*, *B. fragilis*, sometimes *Cl. perfringens.*

DIAGNOSIS

A specimen of peritoneal exudate is examined in the same way as pus from a wound.

TREATMENT

Appropriate antibiotics: it is usually necessary to start therapy before sensitivity results are available; a suitable combination would be:

Ampicillin
Gentamicin

Metronidazole
Alternative drugs: cefuroxime or cefoxitin
Therapy should be reviewed as soon as the results of sensitivity tests are known.

CLOSTRIDIAL WOUND INFECTION

Wound infection primarily due to clostridia differs clinically from the infections described above in that it is not purulent and therefore is not, strictly speaking, a form of sepsis. Clostridial anaerobic wound infection is rare but much more severe.

The two main diseases associated with clostridial wound infection are shown in Table 34.3

Table 34.3 Clostridial wound infection

Disease	Causal organisms
1. Gas gangrene	*Cl. perfringens (65% of cases)* *Cl. oedematiens* (20-40% of cases) *Cl. septicum* (10-20% of cases)
2. Tetanus	*Cl. tetani*

Gas gangrene

A rare disease in peacetime but a scourge amongst the wounded of armies in the field—notably during the First World War.

CLINICAL FEATURES

A spreading gangrene of the muscles with profound toxaemia and shock: there is oedema with blackening of the tissues and a foul-smelling serous exudate; crepitus (a palpable crackling or bubbling) can often be detected under the skin due to gas production by the clostridia (Plate 22).

Severity: a serious disease with a high case fatality rate often requiring amputation or excision.

PATHOGENESIS

Due to contamination of wounds by dirt and soil containing clostridia derived from animal faeces: infection is favoured by extensive wounding with the presence of necrotic tissue, blood clot and foreign bodies—all of which produce anaerobiasis. Vascular damage may impair blood supply to the site.

Wounds: apart from war injuries, gas gangrene is encountered in civilian practice following major trauma; occasionally it complicates operations on the bowel *(Cl. perfringens* is a normal bowel inhabitant), mid-thigh amputation or vascular surgery on the ischaemic limb.

Clostridia: more than one species is often present in gas gangrene.

Toxins: clostridia produce powerful toxins which themselves cause tissue damage—and so anaerobiasis—and thus enhance spread of the infection.

DIAGNOSIS

Specimen: exudate, tissue.

Direct Gram film: observe for typical Gram-positive bacilli—spores may or may not be seen. In clinical material *Cl. perfringens* is usually capsulated but does not form spores.

Culture: blood agar and aminoglycoside blood agar anaerobically: Robertson's meat medium.

Observe: for typical colonies.

Identification: Cl. perfringens—Nagler reaction; identification of other clostridia may be difficult—it is largely based on biochemical reactions and toxin production.

TREATMENT

Surgical: by wide excision or amputation of affected tissue.

Antibiotics: large doses of penicillin; perhaps with metronidazole in addition.

Antitoxin: widely used in wartime; of doubtful value.

Hyperbaric oxygen to reduce anaerobiasis in tissues has been reported to be effective. This supportive treatment requires special apparatus: if available it should be administered.

Tetanus

A classic example of an exotoxic bacterial disease: due to *Cl. tetani* the toxin produced is one of the most powerful known.

Tetanus is a very rare disease in the UK but the wounds from which it may develop are common and may be trivial.

CLINICAL FEATURES

Severe and painful muscle spasms: the masseter muscles are often affected early in the disease causing 'lockjaw'—the familiar name for tetanus—and 'risus sardonicus'—the characteristic facial grimace produced by spasm of the facial muscles.

Because the extensor muscles of the body are more powerful than the flexor muscles, as the spasms progress the body becomes arched in 'opisthotonus' with only the patient's head and heels touching the bed. Death is due to exhaustion, asphyxiation or intercurrent infection.

PATHOGENESIS

Cl. tetani produces a protein exotoxin—mainly released by bacterial lysis. It has two components:

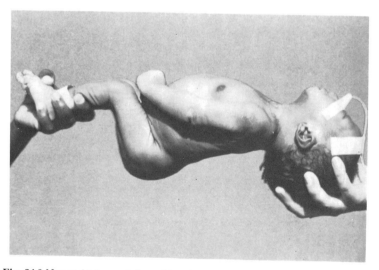

Fig. 34.2 Neonatal tetanus. Infant with opisthotonus due to extensor muscle spasm. (Reproduced with permission from Abbott Laboratories 'Slide Atlas of Infectious Diseases' 1982, Gower Medical Publishing, London. Photograph by Dr. T.F. Sellers Jr.).

1. *Tetanospasmin*, which acts on synapses to block the normal inhibitory mechanism which controls motor nerve impulses.
2. *Tetanolysin:* lyses erythrocytes.

Spread: Cl. tetani does not spread beyond the wound but the toxin, absorbed at the motor nerve endings, travels via the nerves to the anterior horn cells in the spinal cord.

Site: wounds of the face, neck and upper extremities are more dangerous than those of the legs and feet: they are associated with more severe disease and a shorter incubation period (usual range for tetanus is 5-15 days).

Tetanus neonatorum, in which the umbilical stump is the portal of entry, is still common in rural areas of Asia, Africa and South America. Fig. 34.2.

EPIDEMIOLOGY

Source: animal faeces, *Cl. tetani* is hardly ever found in human faeces. Wounds become contaminated with spores and, if anaerobic conditions are present, the spores germinate to produce vegetative bacilli which form toxin.

Worldwide in distribution but the incidence is very much higher in the Third World—where it is an important cause of death—than in the developed countries.

Classically tetanus is associated with severe wounds contaminated with soil or dust similar to those that precede gas gangrene. Today in countries with good medical services, prophylactic measures prevent patients with such wounds developing the disease and tetanus often follows minor injuries disregarded by the patient, e.g. a small penetrating wound from a splinter of wood. In some of the reported cases of tetanus in the UK, no wound could be found.

DIAGNOSIS

The diagnosis is clinical: attempts at bacteriologal confirmation often fail.

Specimen: swab or exudate from wound.

Direct Gram film: examine for characteristic Gram-positive bacilli with round terminal spores—'drum-sticks' (Plate 6).

Culture: blood agar and aminoglycoside blood agar anaerobically; Robertson's meat medium.

Observe: typical translucent spreading colonies; subculture from the edge to obtain a pure growth.

Identify: by Gram film, biochemical tests and demonstration of the exotoxin (provisionally by inhibition of haemolysis on blood agar by specific antitoxin; confirmation by demonstration of mouse pathogenicity and its prevention by antitoxin).

TREATMENT

Supportive: artificial ventilation with muscle relaxants to control spasms.

Antitoxin: large doses intravenously to neutralise toxin.

Excision of wound.

Antibiotics (penicillin or tetracycline): to prevent further toxin production.

PREVENTION

Official policy in Britain is for active immunisation in childhood with formol toxoid as part of the Triple Vaccine.

In casualty departments: a common problem is the wound contaminated with soil or dust and therefore, potentially, with *Cl. tetani*. The wound must first be thoroughly cleansed and then:

(i) If the patient has been previously fully immunised a booster dose of tetanus toxoid should be given; if the wound is dirty and more than 24 h old human antitetanus immunoglobulin must also be administered.

(ii) If the patient is non-immune or if the previous history of active immunisation is uncertain, give human antitetanus immunoglobulin as well as starting a full course of tetanus toxoid by injection at another site.

Patients or the parents of an injured child are often unable to give a history relating to tetanus vaccination. Parents may know that their child has been immunised against the infectious diseases diphtheria and pertussis without realising that the triple vaccine also protects against tetanus.

Penicillin (usually a mixture of short and long acting varieties) has been recommended on its own as prophylaxis but its value is not proven. However, in casualty practice it is common to give penicillin to the wounded not only as part of the scheme to prevent tetanus but also in an attempt to avoid pyogenic infection.

ABSCESSES

A most important form of sepsis—a threat to health and recovery and sometimes very difficult to diagnose. Abscesses, both obvious and

Table 34.4   Abscesses—some common sites

Site	Route of infection— predisposing factors	Bacteria usually responsible
Subcutaneous tissues e.g. finger pulp, palmar space	Penetrating wounds	Staph. aureus
Axilla	Extensions of superficial infection via lymphatics to axillary lymph nodes with suppuration	Staph. aureus
Breast	Breast-feeding—infected from infant	Staph. aureus
Peritonsillar (quinsy)	Streptococcal sore throat	Strep. pyogenes
Intra-abdominal e.g. appendix, subphrenic, paracolic	Appendicitis / Abdominal sepsis, peritonitis due to any cause	Faecal flora, Strep. milleri
Ischiorectal	Direct from rectum	Faecal flora
Perianal	Infected hair follicle round anus	Staph. aureus, faecal flora
Pelvic	Abdominal, gynaecological sepsis	Faecal flora, Genital flora
Tubo-ovarian (pyosalpinx)	Gynaecological sepsis Gonorrhoea	Genital flora, N. gonorrhoeae
Bartholin's gland	Local spread	Genital flora, N. gonorrhoeae
Perinephric	Extension of acute pyelonephritis	Coliforms
Cerebral	Otitis media; sinusitis Haematogenous	Strep. milleri Bacteroides Staph. aureus, Proteus species
Hepatic	Ascending cholangitis Portal pyaemia	Faecal flora, Strep. milleri
Lung	Aspiration pneumonia Bronchial obstruction with collapse (e.g. tumour) Staph. aureus pneumonia	Oro-pharyngeal flora Staph. aureus

Note: the anaerobic members of the commensal flora (Bacteroides species, anaerobic cocci etc.) can also play a role in abscess production; usually in conjunction with other organisms.

cryptic, are an important part of the work of hospital bacteriology laboratories.

An abscess is a collection of pus within and often deeply within the body: abscesses are walled off by a barrier of inflammatory reaction with fibrosis. It is therefore often impossible to treat them satisfactorily by antibiotics alone—surgery and drainage are also necessary.

Abscesses can form in almost any tissue or organ of the body but some sites are much more common than others: these are listed in Table 34.4.

Tuberculosis commonly causes abscesses but these are 'cold' i.e. not accompanied by pain, redness or an acute inflammatory response: they are discussed in more detail in Chapter 38.

CLINICAL FEATURES

Abscesses may be: (i) clinically obvious or (ii) cryptic.

Clinically obvious abscesses include those at sites such as breast, axilla, peritonsillar (quinsy), perianal, ischiorectal and Bartholin's glands: there is painful swelling with local inflammation, fever and often some degree of constitutional upset: brain abscess commonly presents with the signs of a space-occupying lesion most often in the temporal lobe (Fig. 34.3).

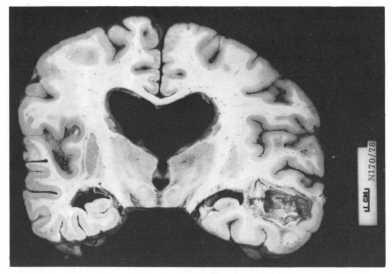

Fig. 34.3 Cerebral abscess. Section of brain with an abscess in the right temporal lobe secondary to suppurative otitis media. (Photograph by Professor J. Hume Adams.)

Cryptic abscesses can be exceedingly difficult to diagnose and, unfortunately, are relatively common: many abdominal and pelvic abscesses are of this type—the patient presents with vague, progressive ill health and fever: although obviously toxic there are no definite localising signs.

In addition to the clinical history there may, however, be other clues: for example, a subphrenic abscess may be detected radiologically: a positive blood culture substantiates sepsis although it does not localise it: CAT scan may help to do this but in some cases, laparotomy is required and should be carried out if there is a high degree of clinical suspicion.

PATHOGENESIS

Abscesses are localised infections but do not necessarily form at the site of primary infection: there may be tracking of pus leading to a purulent collection at a site some distance from the original infection; subphrenic and pelvic abscesses are examples of this.

Metastatic abscesses: occasionally, multiple abscesses form usually as a result of blood-borne or 'pyaemic' spread of infected thrombi; such abscesses are found in many sites. *Portal pyaemia* in which the source is intra-abdominal sepsis (often the appendix) is an example of this which results in liver abscesses.

DIAGNOSIS

Specimen: pus (which may have to be collected at operation or by aspiration); blood culture.

Culture: on blood agar aerobically and anaerobically, Robertson's meat medium; incubation may have to be continued for some days.

Observe: for bacterial colonies or growth.

Identify: by tests appropriate for the bacteria isolated.

TREATMENT

Antibiotics are rarely sufficient on their own: if the abscess does not discharge spontaneously (and this can have serious effects) surgical intervention and drainage may be necessary; this should be done under appropriate antibiotic cover; the drugs to be administered depend on the site of the abscess and, therefore, the likely infecting bacteria (Table 34.4).

35

Pyrexia of unknown origin

Definition. Popularly known as PUO, this clinical entity has been defined in a number of different ways; basically, the patient must have a fever that is:

1. significant (a temperature over 38 °C)
2. persistent (for at least 1 week: usually longer, often 3 weeks)
3. without a readily identifiable cause (i.e. no localising signs or symptoms)

Cause: is more often a relatively common disorder with an atypical presentation or an illness that is notoriously difficult to diagnose than an obscure disease.

Diagnosis: can be exceedingly difficult; in some cases it is never made, and the patient may recover spontaneously.

The principal types of disease which can be responsible for PUO are shown in Table 35.1.

Table 35.1 Causes of pyrexia of unknown origin

Cause		Percentage of cases
Infections		40
Neoplasms	Especially lymphoma, leukaemia, but also other forms of cancer, e.g. hypernephroma, hepatoma, disseminated malignancy	20
Connective tissue diseases	Systemic lupus erythematosis, polyarteritis nodosa, temporal arteritis	20
Others, e.g.	(i) Granulomatous diseases — Crohn's disease, sarcoidosis (ii) Drug-induced fevers (iii) Malingering ('factitious fever')	20

INFECTION

The most important cause of PUO: with a stringent definition,

244

infection accounts for 40 per cent of cases; with less rigid criteria (i.e. lower grade fever of shorter duration) it is responsible for at least 70 per cent of cases.

Bacterial infections which can present as PUO are listed in Table 35.2.

Table 35.2 Bacterial diseases commonly presenting as pyrexia of unknown origin

Cause	Disease—special features
Systemic infections	
Infective endocarditis	Infection of heart valves
Tuberculosis	Pulmonary, non-pulmonary (e.g. bone, renal) or cryptic miliary
Enteric fever	Usually acquired abroad
Brucellosis	Occupational association
Leptospirosis	Occupational association
Localised sepsis	
Hepato-biliary sepsis	Cholecystitis, cholangitis, liver abscess
Intra-abdominal abscess ⎫	Usually follow intestinal or gynaecological
Pelvic abscess ⎭	sepsis; sometimes post-operative
Renal infections	Chronic pyelonephritis; perinephric abscess
Sinusitis	
Dental infection	Apical dental abscess

Other infectious causes of PUO include:

Viral diseases endemic in the UK: infectious mononucleosis (glandular fever), hepatitis A: others such as enterovirus, cytomegalovirus infections, childhood fevers with atypical presentation.

Q fever: due to a rickettsia-like organism, *Coxiella burneti:* most common in males in agriculture.

Psittacosis: due to *Chlamydia psittaci;* suspect if the patient keeps parrots, budgerigars or pigeons.

Protozoal diseases: malaria, amoebiasis (with liver abscess), visceral leishmaniasis (kala-azar), trypanosomiasis—all of which are tropical infections: toxoplasmosis, worldwide in distribution, is usually acquired in this country.

Fungal diseases: histoplasmosis, coccidioidomycosis.

Imported diseases: the common causes of PUO in the UK are very different from those seen abroad. Air travel has allowed the importation of tropical diseases within their incubation period so that they may present in this country as undiagnosed fevers. It is essential to take an accurate 'geographical history' from all patients. Malaria is the most important disease in this category;

almost all cases of typhoid fever in Britain have been acquired abroad.

Lassa fever, Ebola fever and Marburg disease are viral diseases imported from Africa—all rare and very serious; once suspected, patients require to be investigated and nursed in strict isolation because of the danger of infecting nursing and other staff in attendance.

INVESTIGATION

This can be both time-consuming and difficult: below are the main examinations carried out; emphasis has been placed on those that help to establish an infective cause for PUO.

1. *Clinical history:* to include patient's previous illnesses, family illnesses, foreign travel, contact with pets or farm animals, occupation, currently prescribed drugs.
2. *Physical examination:* especially to detect lymphadenopathy or enlargement of liver or spleen.

3. *Laboratory examinations:*
   a. *Bacteriology*
   Blood culture: serial cultures should be taken: isolation of viridans streptococci points to infective endocarditis: isolation of coliforms, *Streptococcus faecalis, Strep. milleri,* or *Bacteroides species* suggests the possibility of intra-abdominal sepsis.
   Urine microscopy and culture: serial specimens may be necessary to detect intermittent bacteriuria in chronic pyelonephritis; microscopic haematuria is often present in infective endocarditis and hypernephroma.
   Stool culture and microscopy: indicated if there is a history of diarrhoea; again several specimens should be examined.
   Serology: appropriate tests for enteric fever (Widal reaction), brucellosis, infectious mononucleosis, Q fever, leptospirosis, and for the other infections listed above.
   b. *Haematology*
   Full blood count: blood films with differential white cell count and examination for malarial parasites: erythrocyte sedimentation rate—a very high result (e.g. more than 100 mm per h.) suggests a connective tissue disease or malignancy.
   c. *Biochemistry*
   Liver function tests, etc.

4. *Radiological examinations:* straight X-rays of chest, abdomen, sinuses and teeth.

Further investigations should be planned at this stage if the diagnosis has not been made. The order in which they are carried out depends on what is considered the most likely cause of the PUO. Examples are listed below:

1. *Serological tests* for connective tissue diseases; e.g. antinuclear factor, rheumatoid factor and other autoantibody tests. Immunoglobulin studies in myeloma.
2. *Specialised radiology and ultrasonography:* to identify the site of disease, e.g. cholecystogram, excretion urogram, barium studies (in inflammatory bowel disease), isotope and, if available, CAT scanning, lymphangiogram (in lymphoma).
3. *Biopsy:* of lymph nodes, bone marrow, muscle, liver, kidney.
4. *Laparotomy:* a last resort if there is evidence of intra-abdominal pathology.

36

Septicaemia and endocarditis

Two different but sometimes interrelated diseases: both are serious infections with a considerable mortality.

SEPTICAEMIA

Literally 'sepsis of the blood': in practice two terms are used to describe the presence of organisms in the blood stream.

1. *Septicaemia:* when the patient shows evidence of sepsis; a severe disease with considerable morbidity and mortality.
2. *Bacteraemia:* term used if the patient shows no signs of clinical sepsis; often asymptomatic and transient, and usually of minor or no clinical relevance.

Septicaemia is of two kinds:

   a. A basic feature of some generalised or 'septicaemic' infectious diseases, e.g. brucellosis, enteric fever.
   b. A complication of more localised infections, e.g. pyelonephritis, peritonitis, cholangitis, pneumonia, osteomyelitis, abscesses of internal organs, etc.; organisms spread from the focus of infection into the blood stream (Table 36.1).

Nowadays the second form is much commoner and is the type that will be described in this chapter: septicaemia as part of generalised infections is described with the diseases concerned.

CAUSAL ORGANISMS

The organisms that cause septicaemia and the underlying infections or associated clinical conditions are shown in Table 36.1. In about 95 per cent of cases a single infecting organism is responsible for an episode of septicaemia.

Note: that infective endocarditis is also associated with septicaemia but is described separately and in more ·detail below.

Table 36.1   Common causes of septicaemia

Predisposing factor	Causal organisms*
Abdominal sepsis (peritonitis, hepatobiliary infection, abscess, etc)	Coliforms *B. fragilis* *Strep. faecalis* *Strep. milleri*
Infected wounds, burns, pressure sores	*Staph. aureus* *Strep. pyogenes* Coliforms *B. fragilis*
Gynaecological sepsis (puerperal infection, pelvic abscess, salpingitis etc).	Coliforms *Strep. faecalis* *Bacteroides species* *Strep. pyogenes* Lancefield group B streptococci
Urinary tract infection	Coliforms *Strep. faecalis*
Osteomyelitis Septic arthritis	*Staph. aureus*
Pneumonia	*Strep. pneumoniae*
Meningitis	*Strep. pneumoniae* *N. meningitidis* *H. influenzae*
Meningitis in neonates	Coliforms Lancefield group B streptococci
Food poisoning	*Salmonella species* (not *S. typhi* or *S. paratyphi*) *Campylobacter fetus/jejuni*
Drip sites, shunts, intravascular catheters	*Staph. aureus* *Staph. epidermidis* Coliforms
Intravenous drug abuse	*Staph. aureus*
Splenectomised patients	*Strep. pneumoniae*
Immunosuppressed patients	Coliforms *Staph. aureus* *Ps. aeruginosa* *Strep. pneumoniae*

*Proportion of septicaemias due to coliforms 40–45 percent, staphylococci 20–25 percent, streptococci 20–25 percent.

CLINICAL FEATURES

Signs and symptoms: variable—sometimes minimal, sometimes severe and rapidly progressive; the predominant signs may be those of the underlying disease (e.g. pneumonia, peritonitis, etc.).

The presenting feature is usually a worsening of the patient's condition with fever, rigors, tachycardia, tachypnoea, cyanosis, hypotension; in elderly patients there may be confusion, agitation or behavioural changes.

Septic shock

Septic shock is a special form of septicaemia in which there is a sudden and catastrophic deterioration in the patient's condition: it is also known as *endotoxic or bacteraemic shock.*

Clinically: shock is characterised by a progressive cardiovascular collapse, often sudden in onset, with severe hypotension and tissue anoxia leading to multiple organ failure (e.g. heart, lungs, liver, kidneys).

Causes: complex; septic shock is usually a complication of septicaemia with Gram-negative bacilli (e.g. after abdominal or urinary tract infection)—and very occasionally of Gram-positive bacterial infection; the mechanism is probably due to circulating bacterial endotoxin which activates the complement system with the release of vasoactive substances and causes intravascular coagulation.

Endotoxin assay: endotoxin can be assayed in the blood by testing for coagulation of an extract prepared from the amoebocytes (blood cells) of the Limulus crab; this test is both specific and extremely sensitive.

DIAGNOSIS

Blood culture

More than one culture may be required: when septicaemia is suspected in a severely ill patient, two or three separate sets of cultures should be taken from different veins at intervals of about 5 minutes: whenever possible before antibiotics are administered.

Observe: for early signs of growth.

Subculture: to blood agar aerobically and anaerobically.

Identify: as appropriate for the organism isolated.

TREATMENT

Control of infection may need surgical intervention (e.g. to drain an abscess, resuture a ruptured viscus, etc.). Antimicrobial therapy is also required and this should be:

1. Bactericidal.
2. Administered intravenously.
3. Prompt and in adequate dosage.

Septic shock needs special resuscitative measures in addition to antibiotic therapy.

Table 36.2 lists some of the antimicrobial drugs of choice: in practice, combinations are often used.

Table 36.2 Antimicrobial therapy in septicaemia

Organism	Antimicrobial drugs
Strep. pneumoniae Strep. pyogenes Strep. milleri N. meningitidis	Penicillin
Staph. aureus Staph. epidermidis Coliforms	Flucloxacillin or vancomycin: often with gentamicin Gentamicin and/or Cefuroxime
Ps. aeruginosa	Gentamicin and/or Azlocillin
Strep. faecalis	Ampicillin
Bacteroides species	Metronidazole

INFECTIVE ENDOCARDITIS

An infection of the endocardium of the heart valves and sometimes of the endocardium around congenital defects.

Infective endocarditis was formerly known as bacterial endocarditis: the change in nomenclature is in recognition of the fact that organisms other than bacteria can cause it.

Mortality: before antibiotics were available the disease was always fatal; even nowadays the case fatality rate is around 20 per cent.

CAUSAL ORGANISMS

These are listed in Table 36.3

CLINICAL FEATURES

There are two clinical forms of the disease:
1. **Acute**
A rapidly progressive disease due to pyogenic bacteria such as

Table 36.3 Causes of infective endocarditis

Organism		Percentage of cases
Bacteria		
Strep. sanguis		
Strep. bovis	viridans or	
Strep. mutans	non-haemolytic	60-70
Strep. mitior	streptococci	
Strep. faecalis		
Staph. aureus		20-30
Staph. epidermidis		
Other bacteria		10
(e.g. corynebacteria,		
Haemophilus species, coliforms)		
Fungi		
Candida albicans		rare
Aspergillus species		
Rickettsia, chlamydia		
Cox. burneti		rare
Chlam. psittaci		

Staphylococcus aureus, Streptococcus pneumoniae or *Strep. pyogenes.*

2. Subacute

The commoner and more chronic disease: also progressive but slowly: due to less pathogenic organisms such as viridans streptococci, *Strep. faecalis, Staph. epidermidis, Coxiella burneti*, etc.

Signs and symptoms: classically—fever, malaise, weight loss, cardiac murmur, anaemia, splinter haemorrhages (i.e. under the finger nails), haematuria, petechiae, splenomegaly.

Nowadays this full-blown syndrome is rare: most patients present with an insidious illness and few or none of the signs listed above: *the laboratory is crucial in making the diagnosis.*

Clinical course: unless adequately treated (and even in some patients apparently so treated) there is progressive damage to the heart valves leading to cardiac failure and death.

PATHOGENESIS

Although most patients have pre-existing cardiac disease, a substantial proportion—as many as one-third—may have had previously normal hearts: in some others, abnormalities are not diagnosed until after infective endocarditis has developed.

Cardiac and other abnormalities which *predispose to infective endocarditis* are listed below:

1. *Rheumatic valvular disease:* e.g. stenosis or incompetence of the mitral and aortic valves following rheumatic fever.
2. *Congenital defects:* e.g. bicuspid aortic valve, septal defects, patent ductus arteriosus, coarctation of the aorta.
3. *Intracardiac prostheses:* replacement of diseased heart valves with prosthetic valves.
4. *Degenerative cardiac disease:* e.g. calcific aortic stenosis.
5. *Drug abuse:* addicts who take drugs intravenously have a high risk of endocarditis—often with atypical clinical features.

Formerly an infection of adolescence and young adult life, the mean age of patients is now over 50 years due to the increasing importance of degenerative disease and the decreasing incidence of rheumatic heart disease.

PATHOLOGY

Thrombi of fibrin and platelets form on damaged endocardium usually of the mitral or aortic valves. As a rule, only a single valve is affected: circulating organisms colonise the thrombi and convert them into infected *vegetations;* the end result is destruction of the valve (Fig. 36.1).

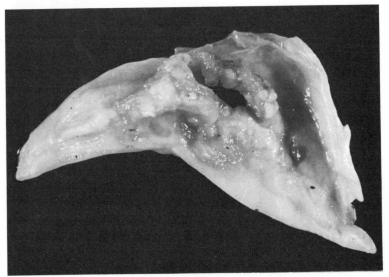

Fig. 36.1 Infective endocarditis. Heart valve from a case of the disease. The valve has a large hole due to the destructive effect of the infection on the tissue of the valve: vegetations can be seen surrounding the area of destruction.

Infection of endocardial thrombi results from:

1. *Transient, asymptomatic bacteraemia:* usually with organisms derived from the normal flora and particularly that of the mouth—which explains the frequency of viridans streptococci as a cause of infective endocarditis; although this type of bacteraemia is common after dental surgery (e.g. extractions), a history of a recent, significant dental procedure is surprisingly uncommon.

2. *Septicaemia:* less common, usually part of a generalised infection with more virulent organisms (such as *Staph. aureus, Strep. pneumoniae*).

Vegetations: shed organisms into the blood stream—often over long periods of time; minute thrombi are also dislodged to give rise to the distant haemorrhagic and other manifestations of the disease.

Immune complexes: can also form and may produce vasculitis and glomerulonephritis.

Prosthetic valve endocarditis

An important complication of cardiac surgery, of *early-onset,* with a high mortality: the infection acquired at operation when it is usually due to staphylococci; *late-onset* endocarditis is acquired in the same way as that of natural valves and although the infecting organisms reflect the normal microbiological pattern of the disease (Table 36.3), staphylococci remain the commonest single cause.

DIAGNOSIS

Blood culture

The corner-stone of diagnosis and important for the subsequent treatment of the patient; repeated cultures may be necessary to isolate the causal organism, e.g. two to six, if possible, taken over 48 h.

Observe: for early signs of growth.

Subculture: to blood agar aerobically and anaerobically.

Identify: as appropriate for the organism isolated.

Antibiotic sensitivity tests

The usual disc diffusion method of testing is not adequate in infective endocarditis. The following tests must be carried out:

Minimum inhibitory concentration of potentially useful drugs both singly and in combination against the organism.

Minimum bactericidal concentration must also be determined since it is important to achieve a bactericidal level of antibiotic in the blood (rather than merely a bacteriostatic level).

Culture negative endocarditis

In 10 to 20 per cent of cases no organisms can be grown from blood cultures. This may be due to:

1. Infection with *Coxiella burneti* or *Chlamydia psittaci*; carry out serological tests to confirm or exclude these diagnoses.
2. Recent antibiotic therapy: repeat blood cultures over a few days in absence of chemotherapy.
3. Infection with fastidious organisms difficult to grow in ordinary media (e.g. bacterial L-forms, *Bacteroides species*): repeat blood cultures using special media if these are suspected.
4. The disease is in a phase in which organisms arc not being shed into the blood stream: repeat blood cultures.

TREATMENT

Depends on adequate—and this usually means high—dosage with an antimicrobial drug; two drugs in combination are often preferred.

Table 36.4 Antibiotic regimens for infective endocarditis

Causal organism	Antimicrobial drugs
Viridans streptococci	Penicillin plus gentamicin
Strep. faecalis	Penicillin* or ampicillin plus gentamicin
Staph. aureus	Flucloxacillin plus gentamicin
Staph. epidermidis	Vancomycin plus gentamicin
Fungi	Amphotericin B plus 5-fluorocytosine
Cox. burneti *Chlam. psittaci*	Tetracycline

* Although *Strep. faecalis* is not sensitive to either penicillin or gentamicin alone, this combination is usually bactericidal and, in practice, effective.
Note: Vancomycin is also a useful drug for infections due to streptococci and *Staph. aureus* especially in patients hypersensitive to the penicillins.

The antibiotic regimen selected must be:
1. Bactericidal.
2. Parenteral (at least initially).
3. Continued for several weeks.

High doses of bactericidal drugs are necessary because the aim of therapy is to eliminate the organisms from their sites enmeshed in the relatively avascular vegetations.

Some of the antibiotic regimens used are shown in Table 36.4.

Laboratory monitoring of therapy. (i) Estimation of the antibiotic level in patient's serum; rarely necessary if large doses are being given intravenously but may be indicated when there is a switch to oral therapy; also used to avoid overdosage with toxic drugs (e.g. aminoglycosides). (ii) Estimation of the bactericidal activity of the patient's serum against the causal organism: although of little prognostic value, a poor result (e.g. no killing at serum dilution of 1 in 8) indicates that the antibiotics chosen, their dose and route of administration should be reconsidered.

Surgery: replacement of damaged valves is now accepted as part of the management of cases of infective endocarditis; it is often lifesaving.

PROPHYLAXIS

Although not of proven value, 'at risk' patients (e.g. those with valvular or congenital heart disease) should be given prophylactic antibiotics—oral amoxycillin or, if hypersensitive to penicillins, oral erythromycin or parenteral vancomycin—before dental procedures. Prior to surgery on the gastrointestinal or urinary tract, parenteral ampicillin with gentamicin is an appropriate prophylactic combination.

Much more importance should be given to improving oral hygiene by encouraging all people to seek better routine dental care.

37

Tuberculosis and leprosy

TUBERCULOSIS

A chronic debilitating disease once the scourge of Victorian Britain ('consumption', 'phthisis',) which remains a major health problem in much of the Third World today.

Causal organisms: Mycobacterium tuberculosis the human tubercle bacillus but also *Mycobacterium bovis* the bovine tubercle bacillus (sometimes regarded as a varient of *Myco. tuberculosis*).

CLINICAL FEATURES

Tuberculosis is a slowly progressive, chronic, glanulomatous infection which most often affects the lungs; other organs and tissues may also be involved.

Respiratory (pulmonary) tuberculosis

Recognised as:
1. Primary infection
2. Post-primary infection.

1. Primary infection

Primary tuberculosis of the lung takes the form of a *primary complex*—a local lesion (Ghon focus) with marked enlargement of the regional hilar lymph nodes. The Ghon focus develops at the lung periphery, usually just below the pleura in the midzone, and is a small lesion.

Many cases of primary tuberculosis are entirely symptomless and the patient, a child or young adult, is unaware of the infection: when symptoms are present, they are often vague and non-specific such as malaise, anorexia, weight loss, fever, sweats, tachycardia; cough may not be prominent.

Progressive primary infection

Disseminated tuberculosis: progressive disease *may* follow primary infection (although it is usually contained by the development of hypersensitivity which causes the primary focus to become walled-off and fibrotic). Antibiotic treatment of primary infection prevents progression but, if untreated, progressive disease is often fatal. Some forms of tuberculosis that result from spread of primary infection are listed below:

1. *Tuberculous bronchopneumonia:* an acute diffuse extension of the infection throughout the lung due to discharge into the bronchial tree of caseous material from an expanding Ghon focus or, more often, from caseous hilar lymph nodes (Fig. 37.1 a and b). *Caseation*—the production of thick cheesy material consisting of pus cells and necrotic tissue—is characteristic of tuberculous inflammatory lesions. A serious and often fatal complication if untreated.

2. *Miliary tuberculosis:* with small tuberculous foci (tubercles) disseminated widely throughout the body as a result of haematogenous (blood-borne) spread of infection.

3. *Tuberculous meningitis:* blood-borne spread of infection to penetrate the blood-brain barrier and involve the meninges; uniformally fatal before the antibiotic era, this is still a serious disease with considerable mortality and disabling sequelae in some survivors.

4. *Bone and joint tuberculosis:* can affect many different sites but a common form is spinal tuberculosis in which there may be collapse of the vertebrae causing kyphosis and formation of a 'cold' psoas abscess in the groin due to tracking of pus down the psoas muscle from the infective process in the spine.

5. *Genitourinary tuberculosis:*

   a. *Renal tuberculosis* presents with frequency and painless haematuria: the urine shows a 'sterile' pyuria—numerous pus cells on microscopy but no growth of pathogens when cultured on standard media.

   b. *Endometrial tuberculosis* in females, *tuberculous epididymitis* in males.

Other sites of primary infection: The primary complex classically involves the lungs but a primary focus with enlargement of the draining lymph nodes may be found elsewhere, e.g. the tonsils with cervical adenitis, the intestine with mesenteric adenitis.

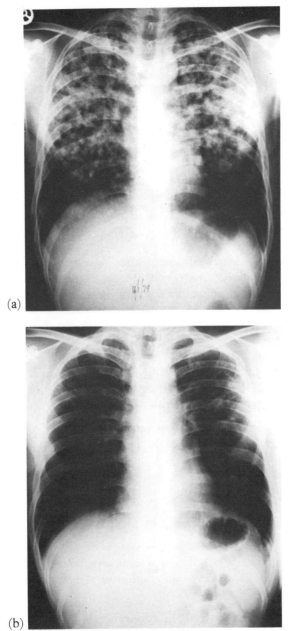

Fig. 37.1 Tuberculosis. (a) Chest X-ray of a patient with tuberculosis bronchopneumonia. (b) Chest X-ray of the same patient 10 months after antituberculosis therapy. (Courtesy of Dr R.S. Kennedy.)

Tuberculosis also causes 'cold' abscesses, i.e. without an acute inflammatory reaction. These abscesses, now rare, most often originate in a cervical lymph gland with discharge of caseous pus and the formation of draining sinuses. Psoas abscess (see above) is another example.

Although these complications often become apparent soon after the primary infection, there may be a latent period of years during which the tubercle bacilli remain dormant before initiating active disease.

2. Post-primary infection

In post-primary tuberculosis the lesions, characteristically more localised and fibrotic, are modified by the development of hypersensitivity by the host.

Post-primary tuberculosis most often involves the lungs with lesions in the apices: if untreated, chronic progressive disease may develop with areas of exudation and caseation surrounded by dense fibrosis; caseous lesions enlarge to coalesce and cavitate: cavities, sometimes large and containing fluid, can then be seen on X-ray: lymph node enlargement is much less marked in this form of tuberculosis.

Presenting symptoms: non-specific ill health with fever: respiratory symptoms include cough, haemoptysis and a pneumonic illness that fails to respond to conventional antibiotics.

The source of post-primary infection can be:
1. Endogenous: reactivation of latent foci formed at the time of primary infection.
2. Exogenous: reinfection by inhalation of infected respiratory secretions from a case of 'open' tuberculosis (i.e. with tubercle bacilli in the sputum).

Today in Britain patients with tuberculosis are usually immigrants or elderly.

Note. Post-primary infection may also disseminate and cause the manifestations listed above as progressive primary infection.

Delayed hypersensitivity in tuberculosis

Tubercle bacilli are readily phagocytosed but then multiply within mononuclear cells and resist digestion: this intracellular parasitism is associated with the development of delayed hypersensitivity (cell-mediated immunity) and of activated macrophages with an increased ability to kill ingested bacilli.

Delayed hypersensitivity modifies the host response to a second challenge with tubercle bacilli and differentiates primary from post-primary lesions. Koch recognised it a century ago when he demonstrated that inoculation of tubercle bacilli into a guinea pig already suffering from a primary lesion with caseating lymph nodes and disseminated infection resulted in only a small superficial lesion at the site of the second inoculation which healed rapidly without lymph node enlargement.

Skin tests in human infection: Delayed hypersensitivity develops in human beings who have been infected, most often without signs or symptoms of disease. It can be demonstrated by the tuberculin test.

Tuberculin test reagents:

1. Old tuberculin (OT)—a crude preparation of the filtrate of a culture of tubercle bacilli: no longer used.
2. Purified Protein Derivative (PPD): obtained by chemical fractionation of OT; PPD is now the preferred material.

Potency of these preparations is assessed in international units and a standard dose is used in tests.

Method of test: all are based on reactions to intradermal inoculation of PPD.

1. *Mantoux:* injection into skin to raise a weal.
2. *Heaf:* multiple intradermal punctures by set of sharp points through a drop of PPD on skin.
3. *Tine:* multiple punctures by kit which also inoculates PPD.

Result: a positive test (indicating that delayed hypersensitivity has developed) produces induration surrounded by erythema at the site of injection after 24 to 48 h.

Interpretation:

1. *Positive* test means degree of immunity (but not complete protection against) tuberculosis. Rarely, a patient with active tuberculosis gives a negative reaction.
2. *Negative* test—depending on the circumstances—is an indication for immunization with BCG vaccine.

EPIDEMIOLOGY

Route of infection mainly by inhalation of infected respiratory secretions, occasionally by ingestion of infected milk.

Reservoirs of infection are patients with 'open phthisis', i.e. tuberculosis in which tubercle bacilli are coughed up in the sputum: there are still in Britain many cases of unsuspected tuberculosis in the general population: elderly people living in poor housing conditions in cities are a particular problem.

Bovine tuberculosis was formerly common in cattle in Britain and resulted in some human cases due to drinking infected milk: the compulsory testing by tuberculin injection of all cattle and the slaughter of positive reactors has succeeded in the virtual eradication of the disease in cows: this measure combined with the *pasteurisation* of milk has resulted in human cases acquired from cattle becoming very rare.

Occupation: some occupations used to be associated with a high mortality rate from tuberculosis: these are trades in which workers are exposed to the inhalation of stone or metal dust such as quarrymen, blasters and tin miners—especially if there is contamination with silica; doctors and nurses also show an increased risk of tuberculosis.

Age incidence: formerly largely a disease of childhood, tuberculosis is now most common in the elderly.

Race: Blacks and American Indians are more susceptible to tuberculosis than Caucasian (white) Americans; environmental or economic factors may contribute to this difference, but it is probably also at least partly due to the severe effects of an infectious disease introduced into a community of people who were not exposed to it until relatively recently.

In the UK the disease is especially common in Asian immigrants: in former days, Highland Scots from rural areas were especially at risk when they moved to the cities.

Genetics: a tendency for tuberculosis to run in families is well recorded: to what extent this is due to an increased risk of exposure is hard to say; twin studies indicate that there may be some degree of increased susceptibility to infection and that this may be inherited.

Immunocompromised patients suffering from disease which affects the functioning of the immune system or on immunosuppressive therapy are extremely susceptible to tuberculosis: in them the infection may be rapid and overwhelming and even in the middle aged, due to depression of delayed hypersensitivity, tuberculosis may behave like a severe primary infection.

DIAGNOSIS

Specimen

These depend on the suspected site of infection.

Respiratory: sputum—if none available, laryngeal swab, bronchial washings or from gastric lavage.

Meningitis: CSF.

Bone and joint: samples removed at operation or by aspiration.

Renal: early morning urine (i.e. the 'overnight' urine voided on waking).

Repeated specimens: three are usually necessary—especially in the case of suspected respiratory and renal tuberculosis.

Direct microscopy

Detection of typical bacilli in a smear stained by *Ziehl/Neelsen* (Plate 7) or auramine, confirms a diagnosis of tuberculosis in most cases. Care is necessary with urinary deposits, however, as these may contain other types of acid-fast bacilli (smegma bacilli).

Culture

All specimens must be cultured: culture is a more sensitive method of detecting *Myco. tuberculosis* than direct demonstration and even if acid-fast bacilli have been detected, culture is necessary for antibiotic sensitivity testing.

Specimens usually require some treatment to remove other bacteria before inoculation of *Lowenstein-Jensen medium.*

Sputum: is 'concentrated' by treatment with sodium hydroxide to liquify the sputum and kill contaminating bacteria; after centrifugation and neutralization with hydrochloric acid, the deposit is cultured.

CSF: is centrifuged before culture.

Bone and joint samples: are disrupted in a sterile grinder before culture.

Urine: is centrifuged and 'concentrated' as for sputum before culture.

Cultures: are examined weekly: *Myco. tuberculosis* and *Myco. bovis* can be distinguished by special laboratory tests including cultural characteristics.

Colonial morphology on Lowenstein-Jensen medium.

1. *Human strains* grow as dry, crumbly yellowish colonies—often described as 'rough, tough and buff' (Plate 8).
2. *Bovine strains* grow as smoother colonies; glycerol does not enhance their growth in contrast to human strains. '

Cultures should not be discarded as negative until they have been incubated for 8 weeks.

Guinea pig inoculation. Guinea pigs are very susceptible to tuberculosis and before adequate culture media had been developed, they were extensively used in diagnosis. Animals injected intramuscularly with clinical material containing tubercle bacilli develop disseminated disease in 4-8 weeks. Rarely require to be used nowadays.

TREATMENT

The discovery of streptomycin in 1944 revolutionised the treatment of tuberculosis. Resistant strains, however, emerge readily and combinations of two or more drugs are used to minimise the risk of this.

Triple therapy: with streptomycin, isoniazid and para-aminosalicylic acid (PAS) was widely and successfully used for treating tuberculosis; it has now been replaced with other regimens.

Streptomycin has to be given by injection: it is also toxic and side-effects include deafness and vertigo (due to damage to the eighth nerve) and nephrotoxicity.

PAS is taken orally but the tablets are large arid unpalatable: nausea and sickness are fairly common complications.

Present recommended therapy

This is a combination of:

Isoniazid   Ethambutol
Rifampicin   Pyrazinamide

These drugs (given orally) are less toxic and better tolerated by patients than the earlier triple therapy. The duration of treatment is now much shorter—6 months if there is no cavitation and 9 months if it is present; ethambutol and pyrazinamide are stopped after 8 weeks.

Alternative or second-line antituberculous drugs:

Streptomycin

Thiacetazone

Para-aminosalicylic acid

Note: tuberculosis was virtually conquered by the introduction of streptomycin in 1948 (one of the most dramatic therapeutic advances in medicine): now largely replaced by less toxic but equally effected drugs (in combination).

The choice of drugs should be reviewed when the sensitivity of the infecting strain to a range of antituberculous drugs has been reported. In patients successfully treated, cultures from the site of infection will soon become negative for tubercle bacilli. If they remain positive, the sensitivity of the isolated organism must be tested again to determine if drug resistance has emerged.

PREVENTION AND CONTROL

BCG vaccine

Consists of live attenuated culture of *Myco. bovis* grown in bile-containing medium: known by the names of the workers who developed it i.e. Bacille Calmette Guérin.

Administered intramuscularly into the deltoid muscle.

Confers hypersensitivity and therefore partial immunity to infection, but prevents the invasive disease characteristic of primary infection.

Age: now in Britain, given between 10 and 13 years of age but only to children who react negatively in Heaf or Tine tests.

Socio-economic factors

Tuberculosis is a disease associated with poverty, malnutrition and overcrowding. Improvement in living standards reduces the incidence: thus deaths from respiratory tuberculosis in the UK fell by two-thirds between 1910 and 1950, i.e. before the introduction of immunisation and chemotherapy.

Identification of infectious cases

Patients with 'open' phthisis are the important source of infection especially in overcrowded conditions; when diagnosed—either by clinical illness or during 'case finding' campaigns—they should be isolated and treated without delay; effective chemotherapy renders nearly all cases non-infectious within a few weeks.

Chemoprophylaxis

Recommended for those at exceptional risk, e.g. children in contact with a known open case e.g. an elderly relative. The drugs chosen are usually isoniazid and rifampicin administered for 6 months.

LEPROSY

The scourge of the ancient world and still afflicting millions of patients mainly in China and India today. Despite the introduction of effective treatment, many, possibly the majority of cases, remain untreated.

CLINICAL FEATURES

Leprosy is a slow, chronic and progressive infection which mainly affects the skin: skin lesions present as nodules or thickened patches and are often associated with thickening of the peripheral nerves and anaesthesia; there may be muscle weakness and ulceration and trophic changes (due to repeated trauma as a result of loss of sensation) in tissues of the extremities leading to the distressing mutilation characteristic of the disease. Lesions mainly affect the skin and exposed cooler extremities such as nose and ears.

Causal organism: Myco. leprae.

Incubation period: long, usually from 3 to 5 years.

Two forms of leprosy are recognised:

1. **Lepromatous:** in which lesions are diffuse and scattered and commonly involve mucous membranes, e.g. of the nose: numerous *Myco. leprae* are present in the lesions; this form of the disease is progressive and severe (Fig. 37.2).

2. **Tuberculoid:** in which the lesions are localised but show early nerve involvement with anaesthesia: there are only scanty *Myco. leprae* in the lesions: in this type of leprosy, lesions tend to be benign and the disease is often self-healing.

Delayed hypersensitivity plays the major part in determining the response of the host to the infection and therefore the clinical type of disease produced: patients with a high degree of hypersen-

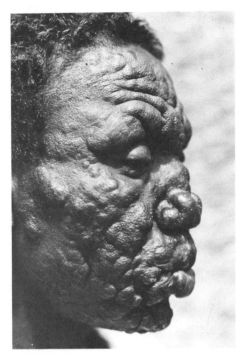

Fig. 37.2 The facial disfiguration of leprosy. This patient has the leonine face sometimes seen in lepromatous leprosy. (Photograph by Dr R. St. C. Barnetson).

sitivity develop the tuberculoid form; in those with no or only a low degree of cell-mediated immunity, the disease tends to be lepromatous.

Mitsuda reaction also called the lepromin test—the intradermal inoculation of an extract of *Myco. leprae* can be used to demonstrate delayed hypersensitivity: a red papule develops at the site of injection in about 48 hours. Positive in tuberculoid leprosy: usually negative in lepromatous leprosy.

EPIDEMIOLOGY

Spread in the community is slow. Leprosy is of *very low infectivity* and even now its mode of spread is uncertain: prolonged contact with cases is probably necessary and the route of infection may be contamination of small cuts and abrasions in the skin; patients with lepromatous leprosy of the nose or skin are probably the main source of infection.

Geographic distribution is widespread (10 million cases worldwide) especially in tropical climates: the disease, however, is found in the southern United States and in some Mediterranean countries in Europe.

DIAGNOSIS

Specimens: skin biopsy or scrapings from lesions in the nasal mucosa.

Direct demonstration of acid-fast bacilli in smears or sections from lesions.

Myco. leprae does not grow on artificial media in the laboratory: it can be cultivated *in vivo* by inoculation of the foot pads of mice and armadillos.

TREATMENT

Dapsone: 4-4—diacetyl-diaminodiphenyl sulphone (DDS) is effective; the drug is unpleasant to take causing side effects such as anaemia, liver dysfunction and skin rashes; treatment must be continued for years and, in lepromatous cases, even for life.

Alternatively, in dapsone-resistant-infections, rifampicin or thiacetazone.

CONTROL

The traditional isolation of patients prevents early cases getting treatment; hospitalisation may be indicated at the start of treatment but thereafter patients regarded as non-infectious can resume their normal activities.

Vaccine: BCG vaccine may offer a degree of protection especially in children.

DISEASES DUE TO ATYPICAL 'ANONYMOUS' MYCOBACTERIA

Atypical mycobacteria are a rather ill-defined group of mycobacteria with growth characteristics, biochemical activity and animal pathogenicity which differ from those of *Myco. tuberculosis:* for example, many produce pigment on culture and grow at lower or sometimes higher temperatures; the pathogenicity and infectivity of these organisms are much lower and in many instances they are present in lung infections as 'fellow travellers' along with *Myco. tuberculosis* and do not play a significant pathogenic role.

Diseases: the principal diseases associated with the different species of atypical mycobacteria are shown in Table 37.1.

Table 37.1 Atypical mycobacteria

Species	Disease
Myco. avium/intracellulare ⎫	
Myco. chelonei ⎬	pulmonary
Myco. fortuitum ⎭	lymphadenopathy
Myco. marinum	abscesses
	swimming pool granuloma
Myco. ulcerans	tropical skin ulcers
Myco. kansasii	pulmonary
Myco. scrofulaceum	cervical adenitis
Myco. xenopi	pulmonary

EPIDEMIOLOGY

Little is known. In 1984 the most common pathogen in the group was *Myco. kansasii.*

TREATMENT

Atypical mycobacteria, although sensitive to rifampicin, are characteristically resistant to other antituberculous drugs, particularly streptomycin and isoniazid. Drug combinations for treatment have to be chosen on the results of laboratory sensitivity tests.

38

Infections of bone and joint

Predominantly infections of childhood: delay in diagnosis and inadequate treatment may result in protracted illness followed by permanent disability.

The two main diseases are:
1. Osteomyelitis
2. Septic arthritis

ACUTE OSTEOMYELITIS

Clinical features: most common in children under 10 years old; dis-

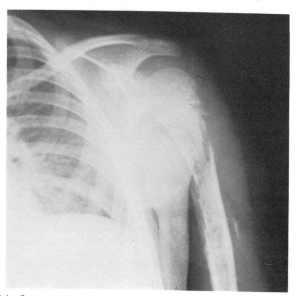

Fig. 38.1 Osteomyelitis of the upper humerus showing bone destruction. (Reproduced with permission from Abbott Laboratories 'Slide Atlas of Infectious Diseases' 1982, Gower Medical Publishing, London. Photograph by Prof. H. Lambert).

tal femur, proximal tibia and proximal humerus (Fig. 38.1) are the classical sites and in adults, the vertebrae. Presentation is bone pain with fever, and local tenderness: the child is reluctant to move the affected limb. There may be a history of preceding mild trauma to the involved bone.

Causal organisms: Staphylococcus aureus (75 per cent or more of cases); other organisms include *Haemophilus influenzae* (in preschool children), *Streptococcus pneumoniae, Strep. pyogenes*. Coliforms or Group B streptococci in neonates. Rarer organisms include salmonellae, *Pseudonomas aeruginosa* (especially in drug addicts) and bacteroides.

Source: not always apparent; septicaemia or sometimes a septic focus elsewhere, e.g. a boil. Spread to bone is haematogenous. Infection at all ages may follow trauma (e.g. compound fracture) or surgical operation.

LABORATORY DIAGNOSIS

Bacteriological diagnosis and antibiotic sensitivity tests are essential for successful management.

1. *Blood culture:* positive in a high proportion of cases; a series of cultures may be necessary.
2. *Culture* of pus collected from the diseased bone either by needle aspiration or at open operation.

TREATMENT

Antibiotics alone are usually effective, especially if started early: surgery is necessary if there is evidence of bone destruction or accumulation of pus.

Antibiotic treatment: initially parenteral, later, the oral route is adequate; in the absence of positive cultures, give flucloxacillin on the assumption that the pathogen is *Staph. aureus*. Alternative drugs include erythromycin, fusidic acid and clindamycin—(the latter two penetrate bone). If a penicillin-sensitive *Staph. aureus* is later isolated penicillin should be substituted for flucloxacillin. Sometimes a combination of antistaphylococcal drugs is preferred.

Strep. pneumoniae: penicillin

Strep. pyogenes: penicillin

H. influenzae: ampicillin

Note. Treatment must be continued for several weeks.

CHRONIC OSTEOMYELITIS

Sometimes due to progression from acute osteomyelitis but may present as a primary infection of bone.

Clinical features: pain, bone destruction with the formation of sequestra and discharging sinuses. If the vertebral column is involved there may be vertebral collapse resulting in paraplegia.

Causal organisms: Staph. aureus—the most common pathogen; rarely *Mycobacterium tuberculosis, Salmonella typhi* (and other salmonellae), *Brucella species.*

LABORATORY DIAGNOSIS

Isolation of the causal agent from:
1. Blood culture.
2. A discharging sinus or material obtained at operation.
3. Possibly from an infective focus elsewhere (e.g. from culture of cold abscess pus in tuberculosis).

TREATMENT

Surgery is usually necessary for the drainage of pus and for the removal of sequestra.

Antibiotics:

1. *Chronic staphylococcal osteomyelitis:* long-term flucloxacillin, fusidic acid or clindamycin; alone or in combination.
2. *Tuberculous osteomyelitis:* isoniazid, ethambutol and rifampicin for 2 years.
3. *Salmonella osteomyelitis:* chloramphenicol or cotrimoxazole.
4. *Brucella osteomyelitis:* tetracycline or cotrimoxazole.

SEPTIC ARTHRITIS

Usually seen as:
1. A complication of septicaemia
2. An extension of osteomyelitis
3. A complication of rheumatoid arthritis
4. Infection following insertion of joint prostheses (see below)

Clinical features: the most striking feature is severe pain which limits movement of the affected joint: in general, only one joint is involved: the onset is sudden with fever, swelling and redness over the joint. Crippling sequelae are common despite antibiotic therapy.

Migratory polyarthralgia and fever are features of rheumatic fever and this differential diagnosis should be considered if there is a history of recent sore throat.

Causal organisms:

> Staph. aureus
> Strep. pneumoniae and other streptococci
> H. influenzae
> Neisseria gonorrhoeae
> N. meningitidis
> Brucella species
> Salmonella species
> Non-sporing anaerobes, e.g. Bacteroides species
> Myco. tuberculosis

LABORATORY DIAGNOSIS

1. *Blood culture*
2. *Examination of fluid aspirated from the joint*
   a. *Direct film:* observe for polymorphs and bacteria; the appearance may allow a presumptive diagnosis and advice regarding immediate chemotherapy
   b. *Culture:* on a variety of media to isolate the causal pathogen.
3. *Culture of specimens from any other infected site,* e.g. throat, genital tract, meninges. If tuberculosis is suspected, examine sputum and urine.
4. *Serological tests* for salmonellosis and brucellosis.

TREATMENT

Antibiotic therapy on a 'best-guess' basis, should be started as soon as diagnostic specimens have been taken. When a causal organism has been isolated, the drug of choice is the same as in osteomyelitis. For arthritis due to *N. gonorrhoeae* or *N. meningitidis,* give penicillin.

REACTIVE ARTHRITIS

Acute arthritis of varying severity and affecting one or more joints which develops 1 to 4 weeks after infection of either genital or gastrointestinal tracts.

Due to an immunological mechanism (c.f. rheumatic fever) and **not directly the result of infection in the joint**: culture of joint exudate is sterile. Two forms are recognised:

1. *Postsexual reactive arthritis* in which the arthritis, usually accompanied by ocular inflammation (conjunctivitis, sometimes also iritis), presents after non-gonococcal urethritis often caused by *Chlamydia trachomatis*. Almost all patients are males: the condition is common in the UK.

2. *Postdysenteric reactive arthritis* in which the arthritis presents after gastrointestinal infection due to shigella, salmonella, campylobacter or yersinia. Affects both men and women: patients may also develop urethritis, conjunctivitis. The condition is common in continental Europe and is now being diagnosed more often in the UK.

Reiter's syndrome: is a term applied to patients who develop the triad of symptoms, arthritis, urethritis and conjunctivitis: much more common in postsexual reactive arthritis.

Predisposing factor: the HLA antigen B 27 is present in 7 per cent of the population but found in more than half of the patients with reactive arthritis. Numerous hypotheses have been proposed to explain this undoubted genetic predisposition but the pathogenic mechanisms that link it with infection and arthritis remain speculative.

INFECTION IN PROSTHETIC JOINTS

The insertion of artificial joints, usually hip or knee, has a high success rate (96 per cent) but failure is usually due to infection. The introduction of inert material predisposes to infection, most often with organisms of low pathogenicity.

CLINICAL FEATURES

The new joint becomes painful with limited movement as a rule soon after operation but sepsis may be delayed for several months.

CAUSAL ORGANISMS

Sometimes *Staph. aureus* but most often skin commensals e.g. *Staph. epidermidis* and corynebacteria.

SOURCE

Contamination of the site at the time of operation with bacteria from:
1. Patient's skin
2. Theatre air
   Late infection may be due to organisms settling in the implant following a transient asymptomatic bacteraemia.

PREVENTION

Careful preparation of the skin site and scrupulous surgical technique: operate either in a specially ventilated theatre or within a plastic isolator tent.
   Antibiotic prophylaxis Peroperative e.g. flucloxacillin alone or with gentamicin.
Note: Laboratory diagnosis of infection may be impossible because specimens are not available. Bacteria in the exudate of the superficial surgical wound are not necessarily the same as those present deep at the site of infection.

TREATMENT

Often must be blind, with flucloxacillin, cephalosporins, fusidic acid; gentamicin may be given in addition.

39

Sexually transmitted diseases

Several infectious diseases are transmitted predominantly or entirely by sexual intercourse. Often the causal organisms are delicate and do not remain viable for long outside the body. Their survival as path-agens therefore depends on transmission by direct contact between mucosal surfaces. Sexually transmitted diseases naturally tend to produce genital lesions but several give rise to systemic, sometimes severe, disease. Generally increasing in incidence possibly due to changing social attitudes, increased travel amongst the young, urbanisation, etc.

Sexually transmitted diseases affect homosexual partners as well as heterosexual relationships. Variations in sexual behaviour can result in these diseases producing lesions in rectum or oropharynx.

The stigma of attending a 'VD clinic' is now less than in former days; increasingly patients with sexually transmitted disease are seen at clinics of genito-urinary medicine; in fact many of the patients are suffering from non-sexually acquired infection and their disease is not evidence of marital infidelity or promiscuity.

The main sexually transmitted diseases are listed in Table 39.1.

GONORRHOEA

A worldwide disease which is still a major problem in the UK.
Causal organism: Neisseria gonorrhoeae.

CLINICAL FEATURES

Acute onset of purulent urethral or vaginal discharge: often with some dysuria and frequency in males; asymptomatic cases are common in females. The disease may involve the rectum or the oropharynx.

Complications: due to local spread—prostatitis, epididymitis in males—rarely urethral stricture in untreated cases, salpingitis in

Table 39.1   Sexually transmitted diseases

Disease	Cause
Gonorrhoea	*Neisseria gonorrhoeae* (the gonococcus)
Non-specific urethritis	*Chlamydia trachomatis*
	? Unknown
Trichomoniasis	*Trichomonas vaginalis*
Vaginal thrush*	*Candida albicans*
Syphilis	*Treponema pallidum*
Vaginitis	*Haemophilus vaginalis, anaerobes*
Genital herpes	Herpes simplex virus type 2
Genital warts	Papilloma virus
Hepatitis B*	Hepatitis B virus
AIDS*	Human T-cell lymphotropic virus III
Chancroid	*Haemophilus ducreyi*
Lymphogranuloma venereum	*Chlamydia trachomatis*
Granuloma inguinale	*Donovania granulomatis* (a Klebsiella-like microorganism)
Pubic lice (crabs)	*Phthirus pubis*
Genital scabies	*Sarcoptes scabei*

*Not always sexually transmitted.
Note: Only gonorrhoea, non-specific urethritis, trichomoniasis, thrush, syphilis and *H. vaginalis* will be dealt with in this chapter.

women; occasionally septicaemia, arthritis, meningitis due to haematogenous spread.

Ophthalmia neonatorum or gonococcal conjunctivitis in the newborn infected during birth from maternal disease.

Gonococcal vulvovaginitis: a rare form of gonorrhoea in young girls either following a sexual offence or sometimes acquired nonsexually by contact with infected exudates or fomites.

DIAGNOSIS

Specimens: urethral, cervical smears and swabs, swabs directly plated or transported to the laboratory in Stuart's or Amies' transport medium.

Direct Gram film: examine for typical intracellular Gram-negative diplococci in smears: often convincingly positive in males but, due to difficulty in interpreting the microscopic appearances in the mixed normal flora, less useful in females (Plate 10).

Culture: on lysed-blood agar medium made selective for *N. gonorrhoeae* by the addition of antibiotics—lincomycin, colistin, amphotericin B and trimethoprim; chocolate agar can be used but the isolation rate is lower. Selective medium is particularly useful in the culture of specimens from sites heavily contaminated with other organisms, e.g. vagina or rectum: incubate in CO_2 for 48 h.

Observe: for typical translucent colonies reacting as oxidase-positive by turning purple on addition of tetramethyl-*p*-phenylene-diamine.

Identify: by acid production (oxidatively) from glucose but not lactrose, maltose or sucrose: alternatively by co-agglutination with monoclonal antibody in a test kit.

TREATMENT

Patients tend to default and whenever possible antibiotics should be given in one curative dose. Numerous different regimens have been proposed.

Standard treatments include:

1. A large intramuscular dose of penicillin or oral ampicillin along with oral probenecid to delay renal excretion.
2. Doxycycline (a tetracycline) in a large oral dose.

Penicillin resistance: normally N. *gonorrhoeae* is extremely sensitive to penicillin but two types of resistance have now been encountered.

   a. *Low-level resistance:* first recognised in 1958: strains remain sensitive to high concentrations of penicillin; relatively common; not due to β-lactamase production.

   b. *High-level resistance:* first recognised in 1976; strains are totally resistant to penicillin; now rapidly increasing in incidence in the UK due to a plasmid-coded β-lactamase.

There are two types of β-lactamase-producing strains one originated from West Africa, the other from South-East Asia. Resistance in African strains is due to a small plasmid which cannot be transmitted to other gonococci. However, Asian organisms contain, in addition to a small plasmid, a second larger plasmid which can *mobilise* the smaller and transfer it to other previously sensitive strains; this has serious epidemiological implications.

Penicillin-resistant gonorrhoea should be treated with cefuroxime, cefotaxime, erythromycin, tetracycline or spectinomycin.

NON-SPECIFIC URETHRITIS OR CERVICITIS

Non-specific genital infection is now the commonest sexually transmitted disease in Britain. Predominantly seen in males—

presumably because infection in females is often symptomless.

Causal organisms(s): almost certainly due to more than one infectious agent but *Chlamydia trachomatis* is the most common identified cause.

Ureoplasma ureolyticus, a mycoplasma, is probably responsible for a small proportion of cases.

CLINICAL FEATURES

Acute purulent urethral discharge indistinguishable clinically from gonorrhoea—but N. *gonorrhoeae* cannot be demonstrated nor isolated: sometimes cervicitis (but this is usually symptomless) in females.

Reiter's disease or syndrome is a triad of urethritis, arthritis and conjunctivitis (with or without iritis) often seen as a complication of non-specific urethritis; sometimes only two of the three symptoms are present.

DIAGNOSIS

Mainly on the basis of clinical signs and symptoms when gonococcal cultures are negative, but attempts should be made to demonstrate chlamydial infection.

Specimens: smears and swabs of urethral or cervical discharge; examine for antigen:

Smears: by indirect immunofluorescence.

Swabs: by ELISA.

Culture: in McCoy or HeLa cell tissue cultures treated with cycloheximide.

Observe: for intracytoplasmic inclusions by Giemsa stain or— better—by immunofluorescence.

TREATMENT

Tetracycline for 7-10 days—but relapses are common.

TRICHOMONIASIS

A.common disease in women: mainly if not entirely transmitted as a result of sexual intercourse with males who have a symptomless infection.

Causal organism: a pear-shaped protozoan—*Trichomonas vaginalis* motile by four flagella (Fig. 39.1).

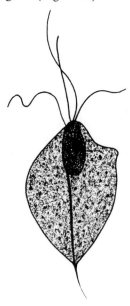

Fig. 39.1 *Trichomonas vaginalis.* This common protozoan is pear-shaped with four flagella.

CLINICAL FEATURES

Vaginal discharge, typically frothy, offensive and usually greenish yellow: there may also be urethral discharge and excoriation of vulva and perineum; the mucosa of vagina and cervix is reddened and inflamed. The bladder may become involved causing dysuria and frequency. Infection is usually symptomatic but varies from an acute severe vaginitis to a mild low-grade or even symptomless infection.

DIAGNOSIS

Specimen swab of vaginal discharge in Stewart's or Amies' transport medium.

Direct wet film: observe for motile protozoa.

Culture: in Fineberg's medium for 5 days: examine for motile protozoa at 2 and 5 days.

TREATMENT

Metronidazole: orally for 7 days or a large single dose. Whenever possible the male partner should be identified, examined and treated.

THRUSH (CANDIDOSIS)

Vaginal thrush is another condition seen in females; genital candidosis is rare in men.

Causal organism: Candida albicans, a normal commensal of mucous membranes including the vagina and bowel. Infection may be endogenous and precipitated by systemic disease (e.g. diabetes) or drug treatment (e.g. broad-spectrum antibiotics); but certainly a proportion of cases are sexually transmitted.

CLINICAL FEATURES

White, membranous patches with itching and irritation of vulva or vagina: white, thick or sometimes watery discharge may be present; many cases are virtually symptomless.

DIAGNOSIS

Can usually be made clinically—with laboratory confirmation.
Specimen: swab.
Direct Gram film: look for characteristic Gram-positive yeast cells with budding pseudohyphae (Plate 15).
Culture: on Sabouraud's medium.
Observe: for typical waxy-surfaced colonies.
Identify: by observation of germ tubes when grown in serum.

TREATMENT

Topical application of nystatin or nitromidazoles (e.g. miconazole).

SYPHILIS

Now a relatively rare disease but important because of its severity and long-term effects if inadequately treated or missed.
Causal organism: the spirochaete *Treponema pallidum.*

CLINICAL FEATURES

There are four clinical stages.

Incubation period: 2-4 weeks.

Primary: a papule usually on the genital area which ulcerates to form the classical chancre of primary syphilis—a flat, dull, red indurated ulcer which exudes serous fluid (Fig. 39.2) this heals spontaneously in 3 to 8 weeks; there is painless enlargement of local lymph nodes.

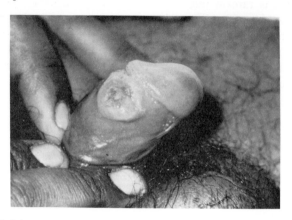

Fig. 39.2 Primary chancre showing indurated ulcer on penis. (Reproduced with permission from Abbott Laboratories 'Slide Atlas of Infectious Diseases' 1982, Gower Medical Publishing, London. Photograph by Dr. R.D. Caterall).

Secondary: 6 to 8 weeks later, the infection becomes generalized with rash—most often papular—of mucous membranes as well as skin: the lesions are highly infectious and contain many treponemes; they may coalesce in intertrigenous areas (especially perianal region) to form wart-like condylomata lata. There is generalised lymphadenopathy in half the patients at this stage and snail-track mucosal ulcers in the mouth in about a third. Other rarer manifestations include periostitis, arthritis, hepatitis, glomerulonephritis and occasionally iridocyclitis and choraido-retinitis.

Tertiary: 3 to 10 years after the primary lesion: *gumma* or granulomatous nodules in skin, mucous membrane or bones: gummata commonly break down to form shallow punched-out ulcers.

Late or quaternary: 10 to 20 years after primary syphilis there are two main clinical forms:

1. *Cardiovascular:* aortitis, aneurysm—classically of the aorta,

aortic incompetence, coronary ostial stenosis.
2. *Neurosyphilis* may take the following forms
   a. *Tabes dorsalis* with characteristic ataxic gait, trophic changes in joints (Charcot's joints): often associated with optic atrophy.
   b. *General paralysis of the insane*, with dementia, tremor, spastic paralysis.
   c. *Meningovascular syphilis*, headache, cranial nerve palsies, pupillary loss of reaction to light (Argyll-Robertson pupil).

Latent syphilis: the disease may lie dormant for many years with no clinical symptoms (but with positive serology). Latent syphilis may eventually develop into cardiovascular or neurosyphilis.

Congenital syphilis

Syphilis is one of the infections—rare amongst bacteria—capable of crossing the placental border to infect the fetus.

Transmission may take place early (10th week) until late in pregnancy: fetal infection is usually seen only with primary or secondary syphilis in the mother.

Congenital syphilis is clinically seen as:
1. *Early*—up to the end of the second year
2. *Late*—after the second year of life
Note: congenital syphilis may be latent with no signs or symptoms
1. *Early:*—generalized infection, low birth weight, skin rash, snuffles, nasal deformity (saddle nose), hepatitis, bone lesions, meningitis, anaemia.
2. *Late:*—interstitial keratitis, bone sclerosis, joint effusions and arthritis, juvenile general paralysis of the insane and tabes: notching of incisor teeth (Hutchinson's incisors), deafness.

DIAGNOSIS

Direct demonstration of spirochaete in fluid or scrapings from chancre or ulcerated secondary lesions by dark-ground microscopy.

Serology: syphilis is unusual amongst bacterial diseases in that it is most effectively diagnosed serologically.

Two types of antigen are used:
1. *Cardiolipin or lipoidal antigen:* not derived from spirochaetes but sensitive in detecting syphilitic antibody, probably directed against cross-reacting cardiolipin in the infecting treponemes.

The antibody disappears with treatment. Biological false positive reactions (BFP) are common.

Tests:
a. *VDRL* (VD Reference Laboratory).
b. (Not used now) Wassermann, Kahn, Meinicke etc.

2. *Specific treponemal antigens* using *Trep. pallidum* as antigen. These tests are specific with fewer BFP reactions: they remain positive after treatment.

Tests
a. *TPHA* (*Trep. pallidum* haemagglutination test)
b. *FTA* (Abs) (Fluorescent treponema antibody absorbed test): this test can be used to detect IgM as well as IgG antibody.
c. *TPI* (*Trep. pallidum* immobilisation test): now infrequently used.

The earliest serological test to become positive is usually the FTA (Abs).

Screening for syphilis on a large scale (e.g. antenatal specimens) is best done by VDRL and TPHA.

Note that other treponemal diseases such as yaws also give positive reactions in all serological lists for syphilis. Table 39.2 lists the typical serological reactions in the different stages of syphilis.

Table 39.2 Serological tests for syphilis

Stage of disease	VD Reference Laboratory (VDRL)	Treponema pallidum haemagglutination (TPHA)	Fluorescent Treponema antibody absorbed (FTA(Abs))
Primary	+ or −	−	+
Late primary	+	+ or −	+
Secondary and tertiary	+	+	+
Late (quaternary)	+	+	+
Latent	+ or −	+	+
Treated syphilis	−	+	+
Congenital syphilis	+	+	+*

*Igm also positive: detection of IgG might only represent passively transferred material antibody.

Note. The success of treatment can be monitored by the VDRL test which becomes negative on successful treatment. The other two tests remain positive.

TREATMENT

Penicillin:

Primary: large doses continued for 15 to 21 days.

Late or latent: large doses for 21 days, usually followed by 10 injections at weekly intervals.

Erythromycin or tetracycline (doxycycline) can be used if the patient is hypersensitive to penicillin; cephaloridine may be effective but some patients show cross-reacting hypersensitivity.

GARDNERELLA VAGINALIS VAGINITIS

Many episodes of vaginitis remain unexplained. *Gardnerella vaginalis*, causes at least some cases of 'non-specific vaginitis or vaginosis' apparently in conjunction with anaerobes such as bacteroides. Infection is probably transmitted sexually although infectivity appears to be low: the bacterium can be isolated from male sexual partners.

Causal organisms: G. vaginalis (previously known as *Corynebacterium vaginale*) a Gram-variable microaerophilic bacillus of uncertain classification; various anaerobes.

CLINICAL FEATURES

Grey-white, thin, non-purulent but very offensive vaginal discharge.

DIAGNOSIS

Microscopy: demonstration of 'clue cells', i.e. squamous epithelial cells with many adherent bacilli.

Isolation of *G. vaginalis* on media such as human blood agar (with production of *β*-haemolytic colonies) or on dextrose starch agar (where the organism produces digestion with zones of clearing round colonies): requires 48 h incubation.

TREATMENT

Metronidazole produces clinical cure—unusual for an infection due to a microaerophilic (i.e. not an obligate anaerobic) microorganism.

CONTROL OF SEXUALLY TRANSMITTED DISEASES

Clearly it is not easy to trace partners of patients with sexually transmitted diseases. In fact the steady increase in the incidence of these diseases in the past two decades illustrates the impossibility of doing so. Nevertheless determined attempts should be made to persuade patients to name consorts and for the consorts to submit themselves to examination and treatment.

40

Infections of the eye

EYELID INFECTIONS

Blepharitis

Red eyelids with fine crusts at the roots of the lashes; may progress to ulceration with destruction of the lids.

Stye

An abscess or small boil in one of the glands of the lash follicles; external styes point to the outer side of the lid margin, internal styes to the inner surface of the eyelid.

Causal organism: Staphylococcus aureus.

Source: endogenous, e.g. from the anterior nares or implantation via the fingers from a septic lesion elsewhere.

Diagnosis: culture of pus or a swab: antibiotic sensitivity tests to topical drugs.

Antibiotic treatment often unnecessary.

CONJUNCTIVITIS

The conjunctival sac is normally colonised by *Staphylococcus epidermidis* and corynebacteria (notably *Corynebacterium xerosis*). Tears act as a defence mechanism by flushing away foreign material; in addition, they contain lysozyme, an enzyme which degrades the peptidoglycan of the bacterial cell wall especially that of Gram-positive organisms.

Conjunctivitis is the result of the invasion of the conjunctival sac by pathogens; children and the elderly are most often affected.

Clinical features: redness of the eye, swollen lids: usually bilateral; the eyes feel hot, gritty and sticky: excessive lacrimation with a mucoid or purulent discharge.

Causal organisms: Staph. aureus, Haemophilus influenzae, Strepto-

coccus pneumoniae, Moraxella lacunata—which classically produces a very red hyperaemic eye, so-called 'angular' conjunctivitis.

Source and spread: sometimes from a stye: often the result of cross-infection via contaminated fingers or fomites. Epidemics, especially in institutions, are common.

Note. Conjunctivitis is often due to adenoviruses.

KERATITIS

Keratitis is inflammation of the cornea.

Clinical features: pain, photophobia, lacrimation and lid spasm: circumcorneal injection: ulceration with grey-white areas and loss of corneal reflex. Severe ulceration is accompanied by iritis and collection of pus at the lower recess of the anterior chamber: may be complicated by corneal perforation leading to panophthalmitis.

Causes: minor trauma, e.g. a foreign body, with secondary infection sometimes due to aspergillus or other soil fungi: may be a complication of conjunctivitis.

Note. Recurrent dendritic ulcer is a serious form of keratitis due to herpes simplex virus.

OPHTHALMIA NEONATORUM

Infection of the baby's eyes acquired either during birth from the maternal genital tract or from some external source after delivery.

Causal organisms:

1. *From the genital tract*
   a. *Neisseria gonorrhoeae—gonococcal ophthalmia:* now rare but formerly an important cause of blindness in children. Clinically, severe purulent conjunctivitis with a tendency to involve the cornea causing ulceration and even perforation. Usually develops within 36-48 h of birth. Can be prevented by instillation of 1 per cent silver nitrate solution into the eyes at birth.
   b. *Chlamydia trachomatis—inclusion conjunctivitis:* clinically, an acute conjunctivitis with mucopurulent exudate: less severe than gonococcal ophthalmia and with a longer incubation period (5-12 days).
   c. *Streptococci of Lancefield group B.*

2. *From an external source*
Staphylococcus aureus—'*sticky eye*': the commonest eye infection, probably spread via medical and nursing staff. Outbreaks of staphylococcal sticky eye in maternity-hospital nurseries are well recognised.

DIAGNOSIS

Bacterial infections

Specimen:

1. Exudate collected directly from the patient's eye with a platinum bacteriological loop: films and cultures should if possible be made at the bedside.
2. Swab of exudate placed in Stuart's transport medium.
   Direct film: examine by Gram's stain for characteristic bacteria.
   Culture: on blood and chocolate agar plates incubated for 48 h at 37°C in the presence of 5-10 per cent carbon dioxide.
   Observe: for growth and identify by usual methods.

Chlamydial infections

Specimen: conjunctival scrapings.
Direct film: examine by immunofluorescence.
Culture: in irradiated or cycloheximide-treated McCoy cells for 2-3 days.
Observe: for intracytoplasmic inclusions by Giemsa stain.

ORBITAL CELLULITIS

A serious infection of the cellular tissues of the orbit which may follow penetrating injury or spread of infection from the nasal sinuses: sometimes 'spontaneous' without any recognised cause. Painful swelling and protrusion of the eyeball are accompanied by general systemic upset. Complications include panophthalmitis, meningitis, brain abscess and cavernous sinus thrombosis—rare nowadays unless treatment inadequate.

Causal bacteria: Staph. aureus, other pyogenic bacteria.

Surgical incision may yield pus—an essential specimen for accurate bacteriological diagnosis.

PANOPHTHALMITIS

Inflammation of the whole substance of the eye which often results in total destruction. Once established in the vitreous, antibiotics fail to penetrate and the eye is rarely salvaged. Usually due to:

1. Spread of infection from a perforated corneal ulcer or orbital cellulitis.
2. Penetrating wounds—traumatic or surgical.
3. Septic emboli

   Causal bacteria:
   a. *Staph. aureus*
   b. *Pseudomonas aeruginosa, Klebsiella species, Proteus species* and *Escherichia coli*

   Diagnosis: culture of a swab taken from the surface of the eye may give misleading results: as soon as possible a sample of pus should be aspirated from the anterior chamber of the eye or from the vitreous by vitrectomy.

TREATMENT

Superficial infections

These usually respond readily to local treatment

1. *General measures:* removal of exudate, irrigation with simple solutions, e.g. saline.
2. *Antibiotics:* either as drops or ointments—chloramphenicol, sulphacetamide, penicillin, aminoglycosides (e.g. framycetin, streptomycin, neomycin), tetracycline, polymyxin and bacitracin. The most widely prescribed are chloramphenicol and framycetin.

Chlamydial infections should be treated orally by tetracycline or erythromycin.

Deep infections

More aggressive antibiotic treatment is necessary:

1. *Systemic antibiotics* unfortunately, penetrate the eye poorly. Concentrations attained in the aqueous exceed those in the vitreous.
2. *Subconjunctival injection:* the route of choice: although painful, up to 1 ml of solution can be injected—penicillin, gentamicin, neomycin, polymyxin can be given in this way.

CHOROIDITIS AND CHORIORETINITIS

Granulomatous reaction in the choroid or retina is usually the result of invasion by microorganisms or parasites: clinically, grey-white choroidoretinal lesions which heal to form scars.

Causal organisms:

1. *Bacteria: Mycobacterium tuberculosis, Brucella abortus, Treponema pallidum* (now very rare).
2. *Viruses:* rubella and cytomegalovirus infection—usually congenital. Varicella-zoster virus (ophthalmic zoster involves the choroid and ciliary body). Herpes simplex virus can cause necrotising retinal vasculitis—a new syndrome.
3. *Fungi: Candida albicans* (retinitis in intravenous drug abusers).
4. *Protozoa: Toxoplasma gondii* (in congenital and acquired toxoplasmosis).
5. *Helminths: Toxocara canis* (mainly retinitis).

41

Zoonoses

Zoonoses are infections between vertebrate animals and man. Not surprisingly, many of those acquired by man affect agricultural workers and veterinary surgeons, but the general public are also at risk for example, through contaminated meat and milk.

Table 41.1 lists the main bacterial zoonoses. Other zoonoses are caused by rickettsiae, viruses and fungi: animal ringworm is perhaps the commonest infection worldwide.

Table 41.1 Bacterial zoonoses

Disease	Causal organism	Main animal host
Food-poisoning	*Salmonella species*	Cattle, poultry
Food-poisoning	*Campylobacter species*	Poultry, other domestic animals
Tuberculosis	*Mycobacterium bovis*	Cattle
Brucellosis	*Brucella abortus*	Cattle
	Brucella melitensis	Goats, sheep
	Brucella suis	Pigs
Anthrax	*Bacillus anthracis*	Cattle
Plague	*Yersina pestis*	Rats
Mesenteric adenitis, enteritis	⌈ *Yersina pseudotuberculosis* ⌊ *Yersina enterocolitica*	Various animals
Septic dog bite	*Pasteurella multocida*	Dogs and cats
Tularaemia	*Francisella tularensis*	Squirrels, rodents
Leptospirosis	*Leptospira interrogans*	Rats, pigs, dogs, cattle
Listeriosis	*Listeria monocytogenes*	Various domestic and wild animals
Erysipeloid	*Erysipelothrix rhusiopathiae*	Pigs, other animals and fish
Rat bite fever	⌈ *Streptobacillus moniliformis* ⌊ *Spirillum minus*	Rats, mice

Domestic animals are more likely to be sources of infection than wild animals due to their closer contact with people. Table 41.2 shows the

Table 41.2 Domestic animals: sources and routes of infection

Source	Route	At risk
Infected animals	Contact	Farm workers, veterinary surgeons, slaughtermen: may be associated with injury.
Contaminated pastures, straw, dust, soil	Inhalation, contact	Farm workers, veterinary surgeons
Milk	Ingestion	People drinking unpasteurised milk*
Meat	Ingestion	Anyone eating meat
Hides, bones, other animal products	Contact	Industrial workers handling animal products, occasionally general public
	Inhalation	Industrial workers handling animal products

*Almost all milk sold in the UK is now heat treated. Pasteurisation became compulsory throughout the UK in 1985 although direct sales of raw milk from the farm are still permitted in England and Wales but not in Scotland. Before these regulations were introduced milk was the vehicle for outbreaks of a number of infections especially salmonellosis and campylobacter enteritis.

principal sources and routes of infection acquired from domestic animals.

Salmonella and campylobacter food-poisoning are by far the most common zoonoses in Britain: these diseases and tuberculosis due to *Mycobacterium bovis* are described in Chapters 29 and 37 respectively.

BRUCELLOSIS

Although brucellosis in Britain is becoming a rare disease, it remains worldwide in distribution: the causal *Brucella species* are named after Bruce who discovered the cause of one form of the disease while serving as an army doctor in Malta.

Causal organisms—brucellae—are small Gram-negative coccobacilli: the main species are listed in Table 41.3.

Table 41.3 Brucella species

Species	Animal host	Geographical distribution
Brucella abortus	Cattle	Worldwide
Brucella melitensis	Goats and sheep	Mediterranean area
Brucella suis	Pigs	USA, Denmark

CLINICAL FEATURES

Incubation period: 1 to 3 weeks, occasionally several months.

Signs and symptoms: undulant fever, a prolonged debilitating febrile illness with remissions and relapses, often becoming chronic when it persists for months or even years: sweating, anorexia, constipation, rigors, weakness and lassitude are the main symptoms; the spleen and lymph nodes are often enlarged, there may be arthritis, orchitis and neuralgia; acute brucellosis is a *septicaemic* illness: abortion is not a feature of human brucellosis.

In chronic brucellosis the symptoms are of vague ill-health: 'psychiatric morbidity' (usually depression) may overshadow physical symptoms.

Duration: on average 3 months: even without therapy symptoms of acute brucellosis usually disappear within one year.

Severity: brucellosis due to *Br. melitensis* tends to be a more severe disease than that due to *Br. abortus.*

Mortality: even before antibiotics, brucellosis had a low case fatality rate—around 2 per cent.

Brucella suis infection is less common than that due to *Br. abortus* or *Br. melitensis.*

PATHOGENESIS

Route of infection: usually by drinking unpasteurised *contaminated milk:* farm workers, slaughtermen and veterinary surgeons are not uncommonly infected by *direct contact* with infected animals or their products—especially the placenta or uterine discharges from parturient animals.

Spread in the body is via lymph channels to lymph nodes and blood stream: the organism then becomes widely distributed in organs and tissues to produce the symptoms of acute brucellosis.

Intracellular parasitism: brucellae have a particular tendency to persist intracellularly notably in spleen, liver and lymph nodes; this is the reason for the well-known difficulty of eradicating the infection by antibiotic therapy. Antigen release from these sites may lead to an immune complex syndrome and cause the symptoms of chronic brucellosis.

DIAGNOSIS

Isolation

Specimens: blood culture in liver infusion or glucose serum broth incubated in CO_2.

Observe for growth: retain cultures for six weeks.

Identify: by dye test, H_2S production, urease activity and agglutination with monospecific antisera.

Positive blood cultures make certain the diagnosis of acute brucellosis but numerous sets should be taken because the organism is notoriously difficult to isolate. *Br. melitensis* is more readily cultured than *Br. abortus*.

Serology

Detect antibody levels against both *Br. abortus* and *Br. melitensis*. Confusingly, some biotypes of *Br. abortus* have a preponderance of melitensis antigen with a minor content of abortus antigen: infection with *Br. abortus* may therefore result in higher titres of antibody to *Br. melitensis* than to *Br. abortus*.

1. *Direct (standard) agglutination test.* Dilutions of the patient's serum are tested against a brucella suspension and examined for agglutination: *prozones* (i.e. absence of agglutination in low dilutions of serum which contain high levels of antibody) are common.
2. *Indirect agglutination test.* Incomplete antibodies which combine with the organisms in the brucella suspension but are unable to agglutinate them may be found in brucellosis: detect by *Coombs' test.*

   Coombs' test: tubes from a negative direct agglutination test are centrifuged to deposit the bacteria; antihuman globulin is added and if incomplete antibody is present which has coated the bacteria agglutination will now take place.
3. *Complement fixation test:* useful in detecting incomplete antibody: widely used because easier to perform than Coombs' test.
4. *Radioimmunoassay:* not generally available but the most sensitive and best test; the only test that can distinguish IgM, IgG and IgA brucella-specific antibodies with certainty.

Table 41.4 Tests for Brucella antibodies

	Direct agglutination test	Indirect agglutination test	Complement fixation test
Immunoglobulin class responsible for a positive test	Mainly IgM but also IgG and IgA	Mainly IgG but also IgA and IgM	Mainly IgG but also IgM *not* IgA
Significant titre	⩾80	⩾80	⩾16

Immunoglobulins involved in the main serological tests for brucellosis are shown above together with the titres regarded as significant (Table 41.4).

Acute brucellosis: at the time of presentation antibody levels are almost always high—direct agglutination titres of 1000 or more are common: IgM and IgG are raised and titres decline slowly to low levels or zero with clinical recovery.

Chronic brucellosis: levels of IgG and IgA but not IgM are elevated: the indirect agglutination test and the complement-fixation test may be positive and the direct agglutination test negative.

Similar findings are encountered in healthy individuals whose immunity is being repeatedly stimulated by contact with brucella organisms and 'positive' serological tests are common in those at occupational risk.

Note. It is impossible to distinguish by serological tests chronic active brucellosis from seropositive but inactive infection—especially in rural areas where exposure to infection is common.

The administration of tetracycline or cotrimoxazole (a 'therapeutic test') may result in clinical improvement and a fall in antibody levels.

Delayed hypersensitivity: the *brucellin* skin test is similar to the tuberculin test and indicates present or past infection with brucellae: often positive in healthy people from agricultural areas. Of little diagnostic value—in fact, the test may stimulate antibody production and make assessment of serological tests even more difficult.

TREATMENT

Acute brucellosis should be cured if appropriate treatment is given promptly: the outcome in chronic brucellosis is much less favourable. *Tetracyclines:* but prolonged treatment is necessary, e.g. for 4-6 weeks: *streptomycin* may be given in addition for three weeks of the course and this combined therapy is probably superior.

Cotrimoxazole: the alternative treatment: usually effective if given for 6 weeks.

BRUCELLOSIS IN ANIMALS

Animal brucellosis is also a chronic debilitating, septicaemic disease: animals are particularly infectious at parturition because of the heavy contamination of placenta and products of conception. *Erythritol,* which is contained in placental tissue, is a growth factor for brucellae.

Abortion is a common sequel of brucellosis in cattle and can result in serious economic losses in herds.

Milk: brucellae are usually excreted in the milk of infected animals.

CONTROL

Brucellosis was eradicated in cattle in Britain in 1981.

Vaccination: of calves between the 4th and 6th month with a live attenuated vaccine (S19)results in significant protection from, but not the elimination of, infection and abortion. In Britain vaccination was abandoned in favour of eradication.

Eradication

1. *Infected herds* are recognised by detecting antibodies to *Br. abortus* in pooled milk samples. The milk ring test is used.
   Method: add a concentrated suspension of *Br. abortus* stained with haematoxylin to the milk and centrifuge.
   Observe: in a positive test the development of a deep blue ring in the cream layer.
   Mechanism: antibodies agglutinate the stained bacteria which rise to the surface with the fat globules.
2. *Individual infected animals* are then identified by isolating *Br. abortus* directly from cream or demonstrating antibodies in their serum or milk whey: they must then be slaughtered.
   Accredited brucellosis-free herds have been built up by this policy.

HUMAN BRUCELLOSIS IN BRITAIN TODAY

Eradication in animals and the compulsory pasteurisation of milk has led to a dramatic decline in the disease in the UK. In the early 1970s some 600 new cases due to *Br. abortus* were notified each year whereas in 1983, this number had decreased to 12. A significant proportion of these cases and almost all infections due to *Br. melitensis*—a steady 4 or 5 per year—are either contracted abroad or acquired from imported dairy products e.g. cheese made from unpasteurised milk.

ANTHRAX

Anthrax is a disease of animals from which man is only occasionally infected: although a wide variety of animals are susceptible, anthrax is mainly a disease of herbivores and especially cattle.

Causal organism: Bacillus anthracis—a large sporing, capsulated, Gram-positive bacillus.

CLINICAL FEATURES

Cutaneous anthrax

Cutaneous anthrax, or malignant pustule, is due to direct inoculation of the skin from infected animals or animal products: an inflamed but painless lesion with surrounding oedema and a characteristic black eschar: local lymph nodes are usually enlarged: if untreated, it may progress to a septicaemia with death from overwhelming infection. In the 1960s about 11 cases per year were reported in the UK: this figure has now fallen to 3.

Acquired in two ways:
1. *Occupational:* 85 per cent of cases: most often in those associated with the meat trade but also in leather workers, tanners, workers in bone meal factories and in agriculture; malignant pustule of the neck and shoulders was an occupational hazard of hide porters due to rubbing of infected hides carried on their backs.
2. *Non-occupational:* occasionally affects the general public: formerly due to handling of infected shaving brushes, leather goods and clothes: now most commonly seen in amateur gardeners who use bone meal.

Pulmonary anthrax

Also called *woolsorter's disease*—a name indicating its mode of spread by inhalation of spores in workers handling contaminated wool: now a rarity—a disease of the nineteenth century.

Clinically: a severe disease with a high mortality rate: in fact virtually always fatal before antibiotics: characterised by fever, increasing respiratory distress and death.

Gastrointestinal anthrax

Also a lethal disease; due to ingestion of *B. anthracis* or its spores; fortunately very rare.

PATHOGENESIS

Long regarded as a classic example of a disease purely due to the invasive properties of the causal organism, *B. anthracis* is now known to produce its effects also by the formation of three toxins:
1. Oedema factor.
2. Protective factor (antibody to this is responsible for immunity to the disease).
3. Lethal factor.

DIAGNOSIS

Specimens: swab or sample of exudate from malignant pustule: sputum from suspected pulmonary anthrax.
Examination:
1. *Direct film:* Gram stained.
   Observe for typical large Gram-positive bacilli.
2. *Culture:* on ordinary media.
   Observe: for growth of typical 'curled hair lock' colonies.
   Confirm identity of isolated bacilli by subcutaneous inoculation of guinea pigs.
   Observe: local lesion with typical gelatinous oedema and death from septicaemic infection.

TREATMENT

Penicillin: erythromycin in patients hypersensitive to penicillin.

PROPHYLAXIS

The decline in incidence of anthrax in occupational 'at risk' groups is probably at least partly due to the introduction in 1965 of vaccination with an alum-precipitated toxoid prepared from the protective factor.

ANIMAL ANTHRAX

A rapidly fatal, septicaemic infection in which huge numbers of organisms are present in the blood and are shed in discharges from the body orifices.

Pastures used by infected animals become contaminated with anthrax spores and may remain infective for many years.

DIAGNOSIS

Specimen: blood from a sick animal: an ear cut off post-mortem and sent to the laboratory (packed with due precautions).

Examine by:

1. *Direct demonstration* of large Gram-positive bacilli—usually very numerous—in a blood smear: *B. anthracis* does not form spores in the animal body. *McFadyen's reaction:* the demonstration of pink-staining capsular deposits lying between the bacilli in a blood film stained by methylene blue.
2. *Culture* on ordinary media for typical colonies.

CONTROL

Animals suspected of being infected must be notified to the authorities. Dead animals must be disposed of by burning or if this is not possible, by burying in quick-lime; a careful check must be kept on the rest of the herd.

Imported animal products are also subject to inspection and control.

PLAGUE

Plague is a natural disease of rats which occasionally spreads to man via the bite of infected fleas: one of the great epidemic diseases, it was the Black Death (which killed millions in the fourteenth century) and the Great Plague of London and elsewhere in Britain during the mid-seventeenth century. Plague still exists in endemic foci in the Western USA, South America, Africa and the Far East.

Causal organism: Yersinia pestis, a small Gram-negative bacillus.

CLINICAL FEATURES

There are two forms of plague.

1. Bubonic plague

The commoner form; a septicaemic illness with fever, prostration, mental confusion and characteristic enlargement with profuse pus formation of inguinal glands (buboes): high mortality rate—about 50 per cent in untreated cases; not contagious—acquired by the bite of fleas.

2. Pneumonic plague

A rare but highly infectious form of plague affecting the lungs; the route of infection is by inhalation of infected respiratory secretions: virtually always fatal.

Reservoir is mainly rural or sylvatic rats: most human epidemics have been due to spread of the disease to urban rats.

Vector: the rat flea *Xenopsylla cheopis:* plague spreads amongst rats via infected fleas; rat fleas do not usually infest man but do so if the rat population has been killed off by the disease.

Infected fleas transmit the disease during biting: the bacilli multiply in the flea including the proventriculus from where the organisms are injected when the flea bites and sucks blood again.

EPIDEMIOLOGY

The disease may become epidemic in the rat population causing heavy mortality: when there are insufficient rats, the fleas turn to man and spread the disease through human populations.

DIAGNOSIS

Isolation

Specimen: needle aspirate of lymph node (bubo pus): sputum
Gram film: examine for Gram-negative bacilli darker stained at the ends than in the middle (bipolar staining).
Culture: ordinary media.
Identify: biochemical and serological reactions.

TREATMENT

Streptomycin: tetracycline, chloramphenicol and cotrimoxazole are other effective drugs.

CONTROL

The prevention of rats coming ashore from ships: although there seems little danger of epidemic plague nowadays, precautions and sampling regulations remain in force.

INFECTIONS WITH OTHER YERSINIA SPECIES

Yersinia enterocolitica ⎤

Yersinia pseudotuberculosis ⎦ cause enteric infection with mesenteric adenitis (which may mimic appendicitis), diarrhoea and fever.

Rare in the UK but common in some parts of Europe (e.g. Belgium, Scandinavia) and in Canada.

Both organisms are widespread in various animal species: *Y. pseudotuberculosis* causes the disease pseudotuberculosis in animals, but animal infection with *Y. enterocolitica* is usually symptomless.

PASTEURELLA

Pasteurella multocida also known as *P. septica* is a commensal organism of the mouths of dogs and cats: it is a not uncommon cause of sepsis, which is sometimes severe, after bites by domestic animals.

TULARAEMIA

A plague-like disease spread from ground rodents such as squirrels, in USA, USSR and elsewhere: due to *Francisella tularensis*.

LEPTOSPIROSIS

A disease of rodents and sometimes domestic animals transmitted to man by direct or indirect contact with animal urine. Human disease, often due to occupational contact, predominantly affects males and is most common in late summer and autumn. In an average year about 60 cases are diagnosed in the UK but twice that number were recorded in 1983.

Causal organism: Leptospira interrogans. This species is divided into 20 serogroups, three of which, canicola, hebdomadis and icterohaemorrhagiae, are responsible for almost all human cases in Britain.

CLINICAL FEATURES

Incubation period: about 10 days.

An initial *septicaemic phase* lasting 3-7 days which presents as an influenzal illness; the diagnosis is rarely made at this stage. Followed by an *immune phase* when leptospires have disappeared from the blood and antibodies begin to appear, characterised by signs of meningeal irritation (headache, vomiting), conjunctival suffusion and evidence of renal involvement (proteinuria). Leptospirosis is a recognized cause of *aseptic (lymphocytic) meningitis.*

Severe leptospirosis *(Weil's disease),* associated with jaundice, haemorrhages and serious renal damage is uncommon and usually due to icterohaemorrhagiae infections. It is likely that many mild cases of leptospirosis are never diagnosed.

SOURCES OF INFECTION

Canicola infections

Natural host: pigs and dogs.
At risk: those who tend pigs and dogs. The infection is now less common in dogs because many are immunised against the disease. This serogroup responsible for less than 10 per cent of human infections.

Hebdomadis infections

Natural host: field mouse; cattle are secondarily infected from pasture or fodder contaminated with mouse urine.
At risk: farmers, often infected from cow urine during milking: the hebdomadis serogroup was the commonest recognised cause of leptospirosis in Britain in 1983 and in before that had been responsible for almost as many cases as the icterohaemorrhagiae serogroup.

Icterohaemorrhagiae infections

Natural host: brown rat.
At risk: agricultural workers, fish handlers, miners, sewage workers and others subject to occupational exposure by employment in a moist environment contaminated with rat urine: also water-sport enthusiasts.

Farm workers are now the occupational group most commonly affected by leptospirosis in the UK: formerly it was a disease of sewage workers and miners but in recent years they have been protected by pest-control measures and safety clothing.

Entry of the pathogen is through skin cuts and abrasions; sometimes via the nasopharynx or conjunctiva following bathing or accidental immersion in infected water, usually stagnant ponds or canals.

DIAGNOSIS

Isolation

Difficult and rarely accomplished.
Specimens:
Blood during the first week of illness.
Urine (the sample *must* be fresh) during the second and third weeks.

Inoculate and observe:

(i) a suitable serum-enriched liquid medium—growth of leptospires.
(ii) guinea-pigs intraperitoneally—jaundice and death.

Serology

The usual method of diagnosis: antibodies are not present in the first week of illness but as a rule can be detected in the second and third weeks.
1. *Agglutination* or *complement fixation test* with a genus-specific antigen (e.g. the saprophytic patoc strain of *L. biflexa*) can detect antibodies to any of the serogroups: if this screening test is positive, carry out the Schuffner test.
2. *Agglutination-lysis (Schuffner) test:* uses a live suspension of leptospires: serogroup-specific; therefore, dilutions of serum have to be tested against laboratory-maintained cultures of at least the three main British serogroups of leptospires: in Reference Laboratories, the full range of 20 serogroups may be included in the test. A titre of 100 or greater is regarded as significant.

TREATMENT

Penicillin; antibiotics are apparently of limited value in leptospirosis and only modify the disease if given early.

LISTERIOSIS

A rare disease (about 30 cases per year in the UK) although the causal organism is widely distributed in nature—in soil, silage, water and a wide range of animal hosts (cattle, pigs, rodents, birds, fish.)

Causal organism: Listeria monocytogenes, a diphtheroid-like Gram-positive bacillus.

CLINICAL FEATURES

Neonates—infection being acquired as a result of mild or inapparent infection in the mother.

The disease (which may also result in stillbirth) presents in two main forms:
1. Meningo-encephalitis.
2. Septicaemia with enlarged lymph glands and widespread granulo-matous nodules in the viscera—*granulomatosis infantiseptica.*

Adults—particularly if they are immunocompromised develop the disease usually as either a septicaemia or a meningo-encephalitis.

Mild or inapparent infection is probably not uncommon in the general population: in a pregnant woman there is a risk of transmission to the child.

EPIDEMIOLOGY

Method of acquisition of the infection from the environment is obscure. Recent report suggests that a source may be uncooked vegetables contaminated with listeria from animal manure.

DIAGNOSIS

Isolation

Specimens: blood cultures, CSF, swabs from genital tract.

Culture: onto blood agar.

Observe: small colonies surrounded by a narrow zone of β-haemolysis.

Identify: small Gram-positive bacilli like corynebacteria but actively motile when grown in broth at 25 °C.

TREATMENT

Ampicillin; chloramphenicol.

ERYSIPELOID

An inflammatory lesion of the skin usually of the fingers and hand resembling erysipelas but due to a Gram-positive bacillus, *Erysipelothrix rhusiopathiae:* the cause of swine erysipelas but found in other animals, birds and fish.

Occupational hazard of meat and fish handlers, veterinary surgeons.
Treatment: penicillin.

RAT BITE FEVER

Includes two separate rare diseases transmitted to man by rat bites: sometimes seen in laboratory workers handling experimental rats.

Causal organism: either (i) *Streptobacillus moniliformis* or (ii) *Spirillum minus,* a spiral organism.

resembling erysipelas but due to a Gram-positive bacillus, *Erysipe-*

42

Infection in compromised patients

Increased susceptibility to infection in certain clinical conditions has been recognised for a very long time, e.g. tuberculosis and staphylococcal skin sepsis are unduly common in diabetes, and when there is an obstructive element in the pathological process such as a kidney stone or a bronchial carcinoma, infection is the rule.

However, modern medicine has resulted in an increasing number of patients who survive because of advances in the management of their disease: they may present special infection problems to both clinicians and microbiologists.

The microorganisms responsible include recognised pathogens and those considered non-pathogenic in the normal host.

The infections, sometimes called *'opportunistic'* are often unusually severe and may present unfamiliar manifestations.

The patients are unusually susceptible because of some alteration in their normal defence mechanisms against infection: i.e. they are *compromised* either by immunodeficiency or by interference with other body defence mechanisms.

The source of infection is either endogenous or exogenous—the latter often due to organisms acquired from the hospital environment.

IMMUNOCOMPROMISED PATIENTS

Deficiencies of the immune response may affect antibody production, cell-mediated immunity, neutrophil function or combinations of all of these defence mechanisms.

1. Disease

Many diseases depress the immune response. By far the most important are *neoplasms of the lymphoid system:* leukaemia, lymphoma (e.g. Hodgkin's disease), multiple myeloma etc.: solid tumours, even if disseminated, have much less effect. A number of other diseases diminish immunity in a variety of ways: the exact mechanisms involved are often complex and incompletely understood. They include *renal failure, diabetes* and *autoimmune diseases* such as SLE and rheumatoid arthritis.

2. Therapy

Therapy which depresses—or occasionally abolishes—immune function is now widely used. This includes:

a. *Drugs*

   (i) Immunosuppressive drugs
   (ii) Steroids
   (iii) Cytotoxic drugs

   Drugs in these categories, often in combination, are used to treat malignant disease and to prevent graft rejection after organ transplantation. Both the disease itself and the treatment administered predispose to infection.

   Patients with acute leukaemia given aggressive chemotherapy in an attempt to achieve remission often develop severe neutropaenia (neutrophil count less than 500 per μl) and are at special risk of septicaemia, usually with coliforms, which may be rapidly fatal. Infection is the cause of death in 75 per cent of patients and half of those who die are septicaemic.

   Patients after transplantation of kidneys, bone marrow or other organs are maintained for long periods of time on an immunosuppressive regimen specifically designed to reduce the cell-mediated immune response which causes graft rejection. Infection is a particular problem in these patients and the most common cause of death.

b. *Radiotherapy*

c. *Splenectomy* results in increased susceptibility to infection with encapsulated bacteria especially *Streptococcus pneumoniae*.

3. Congenital

Rarely, children are born with congenital deficiency of the immune system: this may involve:

a. *Immunoglobulin synthesis:* e.g. B-cell deficiency with depressed production of immunoglobulins. All immunoglobulins may be affected as in the Bruton type of hypogammaglobulinaemia or only some, as in hereditary telangiectasia with deficient IgA and IgE.

b. *Cell-mediated immunity:* T-cell deficiency e.g. thymic hypoplasia (Di George's syndrome).

c. *Combined immunodeficiency:* lack of differentiation of the common lymphoid stem cell resulting in both B- and T-cell deficiency, e.g. Swiss-type agammaglobulinaemia in which there are no lymphocytes or plasma cells in lymphoid organs and the thymus is very small.

Table 42.1   Causes and results of infection in immunocompromised patients

Type of infectious agent	Main microorganisms involved	Clinical manifestations of infection
Bacteria	*Escherichia coli* *Klebsiella species* *Pseudomonas aeruginosa* and other coliforms	Urinary infections, sepsis of colonic origin (e.g. ischiorectal abscess), pneumonia, septicaemia, meningitis
	Legionella pneumophila	Pneumonia
	Mycobacterium tuberculosis	Pulmonary, miliary tuberculosis
	Mycobacterium avium-intracellulare	Pulmonary, disseminated disease
	Staphylococcus aureus	Soft tissue sepsis, pneumonia, septicaemia
	Streptococcus pneumoniae	Pneumonia, septicaemia, meningitis
	Listeria monocytogenes	Septicaemia, meningitis, arthritis
	Nocardia asteroides	Pneumonia; metastatic abscesses, e.g. brain
Fungi	*Candida albicans*	Local thrush, systemic candidosis
	Cryptococcus neoformans	Meningoencephalitis
	Aspergillus; especially *A. fumigatus* *Mucor species*	Pulmonary, occasionally disseminated infections
Viruses	Herpes simplex virus	Severe cold sores
	Varicella-zoster virus	Zoster, sometimes generalised zoster
	Cytomegalovirus	Pneumonitis
	JC (human polyoma) virus	Progressive multifocal leucoencephalopathy
	Papilloma virus	Warts (often florid)
Protozoa	*Toxoplasma gondii*	Severe toxoplasmosis with involvement of retina and brain
	Pneumocystis carinii	Interstitial pneumonia
	Cryptosporidium	Chronic diarrhoea; malabsorption

d. *Neutrophil function:* several syndromes are recognised which affect different aspects of phagocytosis, e.g. chronic granulomatous disease, Chediak-Higashi syndrome.

Table 42.1 shows the principal causes and results of infection in immunocompromised patients.

Source of infection

Endogenous infection: caused by microorganisms that are part of the normal flora, e.g. septicaemia from colonic bacteria, thrush from candida in the mouth.

Exogenous infection: acquired from the environment.

Note: an exogenous potential pathogen, often an antibiotic-resistant hospital strain, may colonise the patient and be assumed as part of the normal flora before causing infection.

Some infections may be either newly acquired or the result of reactivation of asymptomatic latent infection, e.g. tuberculosis, toxoplasmosis, pneumocystis pneumonia, infections due to herpes viruses. *Transplanted tissue* may be the source of a variety of infections, e.g. primary cytomegalovirus infection acquired by a sero-negative recipient from a sero-positive donor.

Prevention of infection

1. *Surveillance:*
   a. Careful clinical examination to detect infection early: institute treatment without delay.
   b. Screening for colonisation by potential pathogens: take repeated cultures from a variety of body sites—of debatable value.
2. *Antibiotics:* avoid indiscriminate use of 'prophylactic' broad spectrum antibiotics: this promotes an abnormal flora of resistant bacteria. However, the bacterial load of aerobic Gram-negative bacilli in the colon can be reduced without predisposing to superinfection by giving,
   a. Oral non-absorbable antibiotics, e.g. FRACON (framycetin, colistin, nystatin) or GVN (gentamicin, vancomycin, nystatin).
   b. Oral cotrimoxazole: also results in selective decontamination of the colon with a consequent reduction in infections due to coliforms and, in addition, protects from pneumocystis pneumonia.
3. *Isolation:* particularly indicated if the patient is severely neutropenic: the measures can be either simple (reverse barrier nursing in a single room) or elaborate (nursing in a laminar airflow bed or room and the preparation of sterilised food).

ACQUIRED IMMUNE DEFICIENCY SYNDROME (AIDS)

First recognised in New York and California in 1981 although cases probably date from 1978. By late 1984 approximately 7000 cases had been reported in USA and 600 in Europe—about 100 of these in UK. This 'epidemic' has generated excessive publicity and undue anxiety.

Definition of AIDS

A person with reliably diagnosed disease indicative of immunological deficiency but who has no known underlying cause for that immune deficiency.

Table 42.2 Distribution of disease in AIDS

Disease	Percentage of patients
Opportunistic infection	60
┌*Pneumocystis pneumonia*	*35*┐
└*Others*	*25*┘
Kaposi's sarcoma ·	25
Lymphoma	3
Kaposi's sarcoma & pneumocystis pneumonia	12

Clinical features

1. *Fully developed disease:* patients have opportunistic infections or unusual malignancies or both. (Table 42.2).
   a. *Opportunistic infections* are often recurrent and multiple: *Pneumocystis carinii* pneumonia is by far the most important although evidence of cytomegalovirus infection is found in almost all patients. A wide variety of other opportunistic infections have been recorded notably due to herpes simplex virus, *Candida albicans, Cryptococcus neoformans, Mycobacterium avium-intracellulare, Mycobacterium tuberculosis, Salmonella species, Toxoplasma gondii* and cryptosporidium.
   b. *Malignancies:* Kaposi's sarcoma is the most common tumour: classically an indolent skin neoplasm associated with African males, in AIDS patients it is aggressively malignant and involves the viscera. The other tumours are lymphomas and these include Burkitt-like lymphoma, non-Hodgkins lymphoma, isolated cerebral lymphoma, angioblastic lymphadenopathy and lymphoblastic preleukaemia.
2. *Progressive generalised lymphadenopathy* (PGL)—(AIDS related complex (ARC), little AIDS, pre-AIDS)—patients are ill with fever and sweating, wasting, generalised lymphadenopathy, oral candidosis, chronic diarrhoea: in an uncertain proportion the disease will progress to full blown AIDS, the remainder seem to recover.

Victims of the disease (either Aids or PGL)

These usually belong to well defined 'at risk' groups (Table 42.3). A disease of males (half in the age range 30–39 years), but women can acquire infection from a partner who is bisexual and babies have been infected *in utero*.

Table 42.3 Distribution of cases by risk group

Risk group	Percentage of cases	
	USA	Europe
Male homosexual or bisexual*	73	77
Intravenous drug abusers	17	3
Haitians	4	3
Africans	–	7
Haemophiliacs	1	3
Blood transfusion recipients	1	1
Female heterosexual contacts	1	
No obvious risk factor	3	6

*especially in promiscuous homosexuals – average of 60 partners per year.

Origin of the disease

This is obscure but retrospective studies indicate that there may have been cases in Zaire in 1976: from there it seems to have spread to the Caribbean, especially Haiti, later USA and eventually Europe.

Causal organism

Human T-cell lymphotropic virus type III (HTLV III)—a RNA retrovirus: antibody to HTLV III can be demonstrated in almost all patients with AIDS, 90 per cent of those with PGL and 20 to 60 per cent of male homosexuals depending on their degree of risk: antibody is also present in up to 35 per cent of haemophiliacs—related to their source of clotting factor VIII (especially if its origin is American): antibody is absent so far in healthy blood donors in U.K.

The virus can be isolated from the blood, semen and saliva of patients (AIDS and PGL) and from apparently healthy individuals who are antibody positive. Unfortunately, virus has recently been detected in a few antibody-negative at-risk individuals. The proportion of virus carriers at present healthy who will develop the disease later is not yet known.

Routes of infection

1. By intimate sexual contact
2. Intravenously in:
   (a.) Drug abusers

(b.) Recipients of infected blood or blood products prepared from pooled plasma, notably factor VIII.

3. By casual contact, especially if there is percutaneous puncture ('needlestick') with infected blood or secretions. The risk, even to health service staff caring for patients, seems to be very small and the general public is not in danger.

Incubation period

This is 6 to 60 months, based on information from those infected by contaminated blood or blood products.

Mortality

High and progressive: 80 per cent of the earliest-diagnosed cases have now died.

Aetiology of diminished immunity

The *primary* defect produced by the virus is a lymphopaenia due to an absolute reduction in T helper (T4) cells. The number of T suppressor/cytotoxic (T8) cells remains unchanged and as a consequence there is a reversal of the normal T4/T8 subset ratio. This results in a deficient cell-mediated immune response and *secondary* features include decrease in natural killer-cell function, decrease in macrophage phagocytosis, anergy to recall of skin-test antigens and polyclonal B cell activation with hypergammaglobulinaemia although the response to new antigens is decreased.

PATIENTS IN SPECIAL CARE UNITS

The management of these patients often creates unnatural portals of entry for microorganisms by breaching the normal non-specific defence mechanisms.

Indwelling catheterisation of the urinary tract, an old-established procedure, is associated with a very high incidence of infection: the longer the catheter remains *in situ* the more certain it is that infection will develop.

Modern intensive care subjects patients on artificial ventilation to prolonged intubation of the respiratory tract: in many cases their tracheal secretions soon contain large numbers of coliforms (e.g. *Escherichia coli, Klebsiella species, Pseudomonas aeruginosa, Acinetobac-*

ter species) that are not usually associated with respiratory infection. The significance of these organisms in the respiratory tract is often difficult to interpret: sometimes their presence is merely due to abnormal colonisation but on other occasions they cause tracheobronchitis or pneumonia.

An intravenous (or intra-arterial) catheter constitutes a foreign body in an open wound at a site (the skin) which is normally colonised with bacteria. Long-line intravenous catheters to monitor the central venous pressure or to allow total parenteral nutrition often remain in position for many days. The catheter soon develops a fibrin sheath which can become infected with a variety of microorganisms derived either directly from the skin or from bacteraemia by trapping bacteria circulating for short periods of time in the blood: infection of the catheter, usually with *Staphylococcus epidermidis*, may be the source of a septicaemia.

PATIENTS WITH PROSTHESES

In recent years it has become standard surgical practice to implant complex metal and plastic prostheses (i.e. foreign bodies) at sites deep within the patient—usually with dramatic beneficial results: unfortunately · these devices sometimes fail and this is often because of infection.

Orthopaedic prostheses (e.g. hip and knee joints): the bacteria responsible for infection are often skin commensal flora, i.e. *Staph. epidermidis* and/or corynebacteria.

Artificial heart valves: the agents that cause infection on these valves include *Staph. epidermidis, Candida albicans* and other microorganisms not usually associated with infective endocarditis.

43

Infections in hospital

Incidence: 20 per cent of hospital patients suffer from infection: half of them are admitted with their infection i.e. it is *community acquired*—the other half develop their infection during their hospital stay i.e. it is *hospital acquired.*

COMMUNITY-ACQUIRED INFECTION

Commonest are those of the lower respiratory tract (about one third of the total), followed by skin and soft tissues and the urinary tract. Patients with these infections are most often found in paediatric, general medical and geriatric wards.

HOSPITAL-ACQUIRED INFECTIONS

Sometimes called *nosocomial infections. Commonest* are those of the urinary tract (about one third of the total), followed by wounds, lower respiratory tract and skin and soft tissues. Patients with these infections are most likely to be in special care baby units, genito-urinary surgery, orthopaedics, general surgery, gynaecology and geriatrics.

Hospital infection is probably as great a problem today as it was in the pre-antibiotic era. Although antibiotics have reduced mortality, they have failed to alter the incidence of infection. The pathogens responsible for present-day infections are often antibiotic-resistant, especially strains of coliforms and *Staphylococcus aureus.* British hospitals are presently suffering from an outbreak of infection due to strains of *Staph. aureus* resistant to methicillin as well as many other antibiotics.

Age: infection is commonest at extremes of age. Because neonates and the elderly are now more likely to survive and spend a longer time as inpatients, they run a greater risk of becoming infected. They are also less able to resist infection possibly because of less efficient immunity.

Susceptible patients: as a result of advances in medical care, there are now several new 'at risk' groups e.g. the immunocompromised, patients in special care units (e.g. intensive therapy) or with prosthetic implants. A few of the infections in these patients are unusual—for example those due to organisms previously regarded as virtually non-pathogenic or with an atypical, and usually more serious, clinical syndrome than that encountered normally.

The majority of hospital-acquired infections, however, are caused by common organisms e.g. urinary infections due to coliforms, wound infections due to *Staph. aureus* and other pathogens (see p. 231), pneumonia due to *Streptococcus pneumoniae,* food poisoning due to salmonella, *Clostridium perfringens.* Septicaemia is the most serious infection and is associated with significant mortality: it is most often due to coliforms or *Staph.aureus.*

Sporadic (endemic) infections are daily occurrences in hospital and do not usually cause concern unless they threaten the recovery of an individual patient.

Outbreaks of cross-infection are, however, not uncommon. Sometimes they can be traced to a common source e.g. staphylococcal wound sepsis due to a surgeon operating with a boil on his wrist, postoperative panophthalmitis due to antibiotic eye drops in which pseudomonads were growing.

Classical contagious diseases (e.g. measles, chickenpox, infantile gastroenteritis) formerly caused outbreaks in children's wards but these are now rare.

Note: influenza remains a problem in geriatric institutions and epidemics there can cause many deaths.

Sources

Infection may be acquired either:
1. *Endogenously*—with microorganisms from the patients' own normal flora *or*
2. *Exogenously*—with microorganisms from other people or inanimate objects (fomites) in the environment

Exogenous infection:

1. *People* are by far the most important source e.g. members of the hospital staff (medical, nursing, ancillary) or other patients either suffering from infection or asymptomatic carriers able to disperse infection.
2. *Inanimate objects* that can spread infection in hospital include surgical instruments, parenteral fluids, anaesthetic apparatus and ventilators, bed pans and urinals, blankets etc. Within the hospital environment there are other sources such as floors, lockers, baths and toilets, food and water, dust, air-conditioning systems. Potentially pathogenic bacteria on inanimate objects may be spread without increasing in numbers but in the presence of moisture, certain organisms (e.g. pseudomonads) multiply and so enhance the risk of infection.

Spread

The routes of infection are:
1. *Contact:* probably the most important: it may be via
   a. The hands or clothing of staff transmitting microorganisms either from their own body or from other patients i.e. acting as a vehicle for patient-to-patient spread.
   b. Inanimate objects: see above.
2. *Airborne:* this may be via
   a. Droplets of respiratory infection spread by inhalation from person-to-person.
   b. Dust from floors and bedding, exudate dispersed from a wound during dressing, scales shed from skin, to a susceptible site usually a surgical wound. Particularly important with *Staph. aureus.*
   c. Aerosols created by nebulizers, humidifiers, air conditioning systems. Notably concerned with the spread of Gram-negative bacilli (e.g. coliforms, legionella) to the respiratory tract. Patients with endotracheal tubes receiving artificial ventilation are especially at risk.
3. *Ingestion:* outbreaks of food poisoning are not uncommon in hospitals especially where it is difficult to maintain a high standard of hygiene e.g. overcrowded or long-stay psychiatric or geriatric wards. Institutional food often contains large numbers of coliforms and, although their ingestion does not cause gastroenteritis, they can become established in the faecal flora of patients. Reservoirs of

antibiotic-resistant bacteria e.g. klebsiella, pseudomonads are thus created within the ward environment.

Prevention

1. *Education* of medical, nursing and ancillary staff in the basic concepts of infection control. For example:
   a. *Hygiene* in theatres, wards and kitchens: the need to pay special attention to the safe disposal of excreta, soiled dressings etc.
   b. *Good nursing* can limit in an open ward the dissemination of exogenous infection by the contact and faecal-oral routes but cannot contain airborne spread. Can also reduce some endogenous infections e.g. chest-infections in post-operative and unconscious patients.
   c. *Techniques* of theatre asepsis, wound dressing, bladder catheterisation, care of intravenous lines etc.
   d. *Good surgical technique* minimises tissue destruction and haematoma formation, reduces soiling especially during operations on the bowel.
   e. *Frequent hand washing.*
2. *Sterilisation and disinfection:*
   a. Provision of sterile instruments, dressings, surgical drapes etc.
   b. Use of disinfectants in the environment and antiseptics on the skin of patients and the hands of staff.
   c. Increased use of disposable items such as syringes, catheters, tubing etc.
3. *Isolation:* single-bedded rooms (cubicles) are necessary to avoid infection by the airborne route: in addition, strict precautions (e.g. gowns, masks, gloves) for anyone coming into contact with the patient, i.e. barrier nursing.
   a. *Source isolation* prevents spread of infection from an infected patient: requires barrier nursing in a cubicle with extract (negative pressure) ventilation.
   b. *Protective isolation* prevents a susceptible (i.e. immunocompromised) patient being exposed to infection: requires 'reverse' barrier nursing in a cubicle ventilated with sterile (i.e. filtered) air, delivered under positive pressure.
4. *Antibiotics*
   a. *A rational policy* for the use of antibiotics: prophylaxis with carefully-chosen drugs can reduce post-operative infections significantly (see p. 341).
   b. *Prohibit topical application*—which encourages the emergence of

drug resistance—of antibiotics that can be life-saving when given systemically.

5. *Staff health:*

   a. *Exclude* from contact with patients staff suffering from infection e.g. viral respiratory infections, septic lesions.

   b. *Protect staff* by appropriate immunisation e.g. BCG vaccine, hepatitis B vaccine.

6. *Surveillance* of infection within a hospital is of prime importance. Vigilance by all members of staff is essential and information from wards, theatres and laboratories should be collated daily by an Infection Control Team whose key members are the Infection Control Nursing Officer (a specially trained nurse) and the Infection Control Officer (usually a microbiologist).

7. *Administration:* policy making to prevent hospital infection should be the responsibility of a larger group—the Control of Infection Committee—composed of the Infection Control Team together with clinicians, community medicine specialist, pharmacist, administrator etc.

The policy making should ensure, good accommodation (including kitchens, toilet facilities), avoidance of overcrowding, adequate provision of nurses and so on. Administration and medical advisory committees must be made aware of the importance of infection in hospital and how to prevent it.

44

Infections in general practice

Infections are a large part of the general practitioner's workload. Not surprisingly, the diseases seen tend to be somewhat different from those dealt with in hospitals. Although laboratory services are becoming increasingly available to family doctors, most must diagnose and treat patients without laboratory investigation. Antibiotic therapy, therefore, is usually prescribed on a 'best-guess' basis.

RESPIRATORY INFECTIONS

By far the most common infections seen by general practitioners are respiratory. These infections are not only an enormous medical problem but significantly affect the economy due to the large number of days lost from work and also to the considerable cost of the drugs prescribed for their treatment.

The relative frequency of these infections in relation to others commonly encountered in general practice is shown in Table 44.1.

Table 44.1 illustrates the extraordinary frequency of respiratory infection in the community. The incidence of common colds (i.e. coryza) is an underestimate since many patients with them do not consult their doctor: colds, acute bronchitis and more than half the

Table 44.1 Common infections. Their annual frequency as a cause of consultation in a population of 10 000

Infection	No. of persons consulting
Coryza, 'flu-like' symptoms	1700
Acute sore throat	1300
Acute bronchitis	1200
Gastro-enteritis	500
Otitis media	400
Exacerbations of chronic bronchitis	400
Urinary tract infection	250

acute throat infections are due to viruses. Chronic bronchitis is a particular problem in urban industrial areas like Glasgow. Hospital doctors—and microbiologists—may not realise how common are acute infections of the ear, especially in children.

Influenza

The respiratory disease which, in an epidemic year, makes the biggest impact on the community is influenza: epidemics recur every year or two, sometimes annually. The appearance of influenza in the community can always be recognized by a sudden increase in the weekly notifications of absence from work due to sickness and also by a concurrent rise in mortality rate.

OTHER INFECTIONS

Table 44.1 also shows the other common infections dealt with by family doctors. These include infective diarrhoea and urinary infections. Like common colds, many cases of gastroenteritis—and, surprisingly, given the discomfort they cause—urinary tract infections, are never reported to the general practitioner so that their incidence is almost certainly an underestimate.

CHILDHOOD FEVERS

Most of the acute fevers of childhood are viral. Nowadays, their impact on the general population has been greatly reduced by immunization. Ensuring that the recommended schedules are fully implemented is one of the most significant contributions a family doctor can make to the individual's and to the community's health. Varicella,

Table 44.2 Less common infections: rates of consultation in a population of 10 000

Infection	No of persons consulting
Pneumonia	40 per annum
Minor post-puerperal	10 per annum
Pulmonary tuberculosis	1 per annum
Chronic pyelonephritis	1 per annum
Infection of bone or joint	1 every 2 years
Extra-pulmonary tuberculosis	1 every 4 years
Infective endocarditis	1 every 10 years

rubella (vaccination is restricted to adolescent school girls) and mumps (for which a good vaccine is available but is not part of the official schedule of immunization) remain a common cause of consultation in childhood. Children—like the elderly—are seen more often by their doctors than other sections of the population.

RARE INFECTIONS

Table 44.2 lists the incidence of some of the less common infections seen in general practice.

These low figures help to set some of the more severe bacterial infections in proportion in terms of community morbidity: doctors who—like the authors of this book—work only in hospital can easily overestimate their importance and frequency in the community.

COST

Antibiotics: widely prescribed in general practice, they constitute approximately 10 per cent of general practitioner prescriptions. The total annual cost to the Health Service is £90 millon—quite apart from the economic costs referred to above.

INFECTION IN THE COMMUNITY

Despite immunization and effective antimicrobial therapy, infection still has a major impact on the health of the general population. Viruses are largely responsible but bacterial diseases also contribute significantly to community morbidity.

Treatment and prevention of bacterial disease

45

Antimicrobial therapy

Bacterial infections are among the few diseases in medicine for which specific therapy is available. Despite this, infections are still common and treatment may fail.

ANTIBIOTICS

More than a century ago, Pasteur observed that the growth of one microorganism could be inhibited by the products of another. However, most of these early products or 'antibiotics' were toxic to mammalian as well as bacterial cells and were therefore of no therapeutic use. Penicillin, discovered in 1929 but not available for clinical trials until 1940, is the product of a mould, *Penicillium notatum*, and was the first antibiotic drug: it is still the best. Other antibiotics are the products of soil streptomycetes and bacteria of the genus *Bacillus*. Many recently introduced antibiotics, e.g. the semisynthetic penicillins and cephalosporins, have been prepared by the chemical manipulation of existing drugs.

Selective toxicity is the ability to kill or inhibit the growth of a microorganism without harming the cells of the host: an essential requirement for any successful antibiotic.

Chemotherapeutic agents e.g. sulphonamides, trimethoprim and many of the antituberculous drugs, are synthetic drugs with this selective toxicity.

Antimicrobial therapy aims to treat infection with a drug to which the causal microorganism is sensitive. This can be achieved if a sound knowledge of microbiology indicates the most likely pathogen in a given condition and its usual antibiotic sensitivity i.e. on a *'best-guess'* basis: it is better still if the infection is investigated bacteriologically by culture and in vitro sensitivity testing.

Antimicrobial drugs are often classified as *bactericidal* when they kill the infecting bacteria or *bacteriostatic* when they prevent multiplication but do not kill the bacteria: this classification is not always clear cut and may depend on the local concentration of the drug. Apart from a few conditions notably infective endocarditis, appropriate bacteriostatic drugs give excellent therapeutic results.

This chapter deals with the main antibiotics and antimicrobial agents in current use from the point of view of a clinical bacteriologist. Pharmacokinetics are not considered.

The principal drugs are listed in Table 45.1. Subsequent tables summarise information about particular antibiotics including their clinical use: inevitably they are oversimplified and should be used only as a guide to antibiotic therapy.

Table 45.1 Antibiotics and other antimicrobial agents

Drug	Principal use
Penicillins	Wide variety of infections
Cephalosporins	Mainly second-line drugs
Cotrimoxazole	Urinary and chest infections
Aminoglycosides	Severe infections with coliforms
Tetracyclines	Chest infections
Metronidazole	Anaerobic infections
Macrolides	Second line drugs to the penicillins
Chloramphenicol	Typhoid fever, meningitis
Fusidic acid	Staphylococcal infection
Nalidixic acid	Urinary infection
Nitrofurantoin	Urinary infection
Polymyxins	Infections with *Pseudomonas aeruginosa*
Isoniazid	Tuberculosis
Rifampicin	Tuberculosis
Ethambutol	Tuberculosis

Penicillins

Chemical structure:

β-lactam ring

Mode of action: bactericidal; inhibits cell wall synthesis by combining with the transpeptidase responsible for cross-linking of the peptidoglycan; activity depends on an intact β-lactam ring.

Resistance: is common and is due to production by bacteria of an enzyme β-*lactamase* which inactivates penicillin by acting on the β-lactam ring: in many cases bacterial β-lactamases are plasmid-coded.

Antibacterial spectrum: varies depending on individual penicillins:

Table 45.2 The penicillins

Penicillin	Administration	Antibacterial spectrum	Clinical use
Penicillins			
Benzylpenicillin (Penicillin G)	i.m.,i.v.	Gram-positive bacteria Neisseria	Streptococcal, pneumococcal infection; sensitive staphylococcal infection; clostridial infection; meningitis, gonorrhoea, syphilis, anthrax, actinomycosis
Phenoxymethyl penicillin (Penicillin V)	oral		
Aminopenicillins			
Ampicillin	oral, i.m.	Similar to penicillin but in addition, *Strep. faecalis*, almost all *H. influenzae* and many coliforms	Urinary and respiratory infections, enteric fever; in combination with other drugs in severe systemic infections
Amoxycillin	i.v.		
Isoxazolyl penicillins			
Methicillin	i.m.,i.v.	Similar to penicillin but less active drugs. Stable to staphylococcal ß lactamase	Staphylococcal infections
Cloxacillin	oral, i.m.,		
Flucloxacillin	i.v.		
Carboxypenicillins			
Carbenicillin	i.m.,i.v.	Similar to aminopenicillins but in addition, *Ps. aeruginosa* and most *Proteus species* Some activity against bacteroides	Urinary, respiratory, burns and other infections due to sensitive bacteria especially *Ps. aeruginosa;* severe sepsis usually in combination with other drugs
Ticarcillin			
Acylureidopenicillins			
Mezlocillin	i.m.,i.v.	Similar to carboxypenicillins but in addition *Klebsiella species* and greater activity against pseudomonads	As for carboxypenicillins
Azlocillin	i.v.		
Piperacillin	i.m.,i.v.		
Amidinopenicillins			
Mecillinam	oral, i.m., i.v.	Coliforms; low activity against Gram-positive bacteria	Urinary infections; Salmonellosis including enteric fever

i.m. = intramuscular   i.v. = intravenous

Note: in this book, reference to penicillin is taken to mean Penicillin G or Penicillin V

the activity is determined by the side chain of the penicillin nucleus.
The principal penicillins and their antibacterial spectrum are shown
in Table 45.2.

Toxicity: virtually non-toxic; very large doses can therefore be given
if required. Hypersensitivity is a problem both in the form of rashes
(especially with ampicillin) and, rarely, anaphylaxis (with any penicillin given by injection).

Augmentin

Contains: amoxycillin and potassium clavulanate in 2:1 ratio.
Administration: oral
Mode of action: clavulanic acid, a ß-lactam compound without
useful antibacterial activity, is an irreversible inhibitor of most
plasmid-mediated and some chromosomal ß-lactamases. When
combined with a ß-lactamase-susceptible penicillin (e.g. amoxycillin),
it enables the penicillin to resist degradation by ß-lactamases.

Antibacterial spectrum: broad: active against ß-lactamase producing
coliforms, staphylococci and bacteroides.

Clinical use: relatively expensive: reserve for infections of urinary
and respiratory tracts, skin and soft tissues due to amoxycillin-
resistant, augmentin-sensitive bacteria.

Cephalosporins

Chemically similar to the penicillins: the antibacterial activity can be
altered by variation in the side chains of the cephalosporin nucleus.
The drugs are somewhat arbitrarily assigned to three generations: the
early, first generation drugs were introduced in the 1960s and progressive development has culminated in the large number of new third
generation cephalosporins now available.

Chemical structure:

ß-lactam ring

Administration: mostly parenteral although a few oral first generation drugs are available.

Mode of action: bactericidal, similar to penicillin. The cephalosporins are stable to many bacterial β-lactamases and this stability has been increased with the later generations: the inferior antistaphylococcal activity of the newer drugs is due to their less avid binding to the target site.

Clinical use: since most cephalosporins are given parenterally, they are virtually restricted to hospital patients: probably most useful in seriously ill patients especially if infected with more than one susceptible organism. Valuable second-line drugs. Increasingly used as an alternative to an aminoglycoside especially in patients with renal impairment and also for short term peroperative prophylaxis.

Toxicity: low. Rashes, fever, sometimes pain at site of injection. Nephrotoxicity—apparently limited to older drugs especially cephaloridine. Haemorrhage due to increase in bleeding time—probably restricted to latamoxef.

Hypersensitivity: about 10 per cent of people who are hypersensitive to penicillin are also hypersensitive to cephalosporins.

Table 45.3 lists the main cephalosporins, and some of their properties: it is not exhaustive, some drugs not available in the UK are marketed in other countries and numerous other compounds are under development.

Sulphonamides and trimethoprim

Both drugs act sequentially in the synthesis of tetrahydrofolate: widely used in combination because of in vitro evidence of synergism: trimethoprim is now available, like sulphonamides, on its own.

Cotrimoxazole

Contains: sulphamethoxazole and trimethoprim in 5:1 ratio.

Administration: oral, intramuscular, intravenous.

Mode of action: bacteriostatic: sulphonamide competes with para-amino benzoic acid as a substrate for the enzyme dihydropteroate synthetase which catalyses the synthesis of dihydropteroate which is then converted into dihydrofolate: trimethoprim combines with and blocks dihydrofolate reductase which converts dihydrofolate to tetrahydrofolate: these are sequential steps in the synthesis of methyltetrahydrofolate which is required for DNA synthesis.

Table 45.3 The cephalosporins

Cephalosporin	Administration	Antibacterial spectrum	Clinical Use
First generation			
Cephaloridine	i.m., i.v.	Wide range of Gram-positive and Gram-negative bacteria: *Strep. faecalis*, *Ps. aeruginosa*, *H. influenzae* and *Bacteroides species* are resistant *Staph. aureus* is sensitive unless methicillin resistant	Second-line drugs: formerly used in severe sepsis; oral drugs may be of value in 'difficult' urinary infections
Cephalothin	(i.m.), i.v		
Cephalexin	oral		
Cephradine	oral, i.m., i.v.		
Cephazolin	i.m., i.v.		
Second generation			
Cefuroxime	i.m., i.v.	Wide: with marked stability to β-lactamases of Gram-negative as well as Gram-positive bacteria. Active against *H. influenzae* and (especially cefoxitin) *Bacteroides species*. Inferior antistaphylococcal activity	Have largely replaced the first generation drugs in severe systemic infections Widely used in prophylaxis
Cefoxitin	i.m., i.v.		
Cefamandole	i.m., i.v.		
Third generation			
Cefotaxime	i.m., i.v.	Similar to second-generation drugs but in addition activity against *Ps. aeruginosa* (especially ceftazidime)	Similar to second-generation drugs particularly serious sepsis due to susceptible aerobic Gram-negative bacilli
Latamoxef	i.m, i.v.		
Ceftazidime	i.m, i.v.		
Cefsulodin	i.m, i.v.	*Ps. aeruginosa* only	

Resistance: is not uncommon: mainly due to production by bacteria of enzymes resistant to the action of sulphonamide and trimethoprim: the resistant enzymes are plasmid-coded and allow normal metabolism despite blockage of the chromosomal enzymes by the drugs: the gene for trimethoprim resistance is often present on a transposon.

Antibacterial spectrum: broad: active against both Gram-positive and Gram-negative bacteria: *Ps. aeruginosa* is resistant.

Clinical use: widely used for urinary and respiratory tract infections: invasive salmonellosis, pneumocystis pneumonia.

Toxicity: nausea and vomiting; rashes: occasionally thrombocytopenia, leucopenia: mouth ulceration: folate deficiency has been reported.

Sulphonamides

Now rarely used on their own: still prescribed for urinary tract infection (although acquired resistance is a problem) and, in combination with penicillin, in meningitis because of their excellent penetration into the CSF.
Sulphadimidine is the most widely used preparation.

Trimethoprim

Now available on its own: mainly for urinary tract infections but also useful in respiratory infections: the main advantage seems to be that the incidence of side effects will is reduced.

Aminoglycosides

A family of extremely useful antibiotics.
Below are the main aminoglycoside antibiotics:

Aminoglycoside *Main clinical use*
Gentamicin ⎤ Severe infections in hospital due
Tobramycin ⎟ to coliforms; gentamicin
Netilmicin ⎟ is still the most widely used:
Amikacin ⎦ in certain infections usually given in
combination with a β-lactam antibiotic

Streptomycin Now little used: most active
aminoglycoside against mycobacteria

Kanamycin ⎤ May be given orally in 'gut sterilisation'
Neomycin ⎦ regimens prior to surgery, in leukaemia, in chronic
liver disease: neomycin also as topical cream

Administration: intramuscular, intravenous.
Mode of action: bactericidal: inhibition of protein synthesis due to action on bacterial ribosomes: all aminoglycosides cause misreading of messenger RNA to produce amino acid substitution in proteins.
Resistance: is generally due to the acquisition of plasmid-coded inactivating enzymes which cause acetylation, adenylation or phosphorylation of the drugs; amikacin is least affected by these enzymes: sometimes due to decreased transport of aminoglycoside into the bacterial cell and, in the case of streptomycin, may be the result of a

mutation which renders the ribosome insusceptible.

Antibacterial spectrum: coliforms, *Ps. aeruginosa,* staphylococci: streptococci and strict anaerobes are intrinsically resistant.

Toxicity: ototoxicity (vertigo or deafness) and nephrotoxicity are major problems because toxic levels are close to therapeutic levels necessary for treatment. Netilmicin *may* be safer than other aminoglycosides. In patients with renal impairment or on long term therapy, serum levels must be monitored (see p. 44): ensure that the trough level is not excessive (2 mg/1 for gentamicin) and the peak level therapeutic (5–10 mg/1 for gentamicin).

Tetracyclines

Formerly amongst the most widely used antibiotics although there are few specific indications for their use: broad spectrum and remarkably free of serious side-effects, they have been particularly successful in the management of chronic bronchitis.

Tetracyclines:
  Tetracycline
  Chlortetracycline
  Oxytetracycline
  Doxycycline ⎤ more effective; can be given in
  Minocycline ⎦ smaller dosage

Administration: almost always oral.

Mode of action: bacteriostatic: inhibit protein synthesis by preventing the attachment of amino acids to ribosomes.

Resistance: is fairly common: generally plasmid-mediated.

Antibacterial spectrum: broad: both Gram-positive and Gram-negative bacteria: however, some strains of *Step. pyogenes,* pneumococci and *H. influenzae* are now resistant: *Ps. aeruginosa,* and *Proteus species* are intrinsically resistant: active against brucellae, *Mycoplasma pneumoniae,* rickettsia, *Coxiella burneti* (the cause of Q fever), chlamydiae.

Main clinical use: acute exacerbations of chronic bronchitis including long-term prophylaxis: treatment of non-specific urethritis, brucellosis, atypical pneumonia, Q fever, psittacosis.

Toxicity: diarrhoea—due to disturbance of alimentary flora—is common but usually mild and self-limiting: rarely there may be kidney and liver damage—avoid in renal and hepatic failure: contraindicated in pregnancy and in children because the drug may be deposited in the developing teeth with permanent yellow staining and also interfere with bone development.

Metronidazole

Originally used—very successfully—to treat *Trichomonas vaginalis* infections in women: now known to be exceedingly effective against anaerobic bacteria.

Administration: oral, rectal (suppositories), intravenous.

Mode of action: bactericidal: converted by anaerobic bacteria to active (reduced) metabolite with inhibitory action on DNA synthesis.

Resistance: almost unknown.

Antibacterial spectrum: strictly anaerobic bacteria e.g. bacteroides, anaerobic cocci and clostridia: no effect on aerobic organisms: actinomyces are resistant: active against anaerobic protozoa (e.g. *Entamoeba histolytica, Giardia lamblia*).

Clinical use: any anaerobic infection, e.g. abdominal and gynaecological wound sepsis, deep abscesses, peritoneal sepsis: Vincent's angina; peroperative prophylaxis in abdominal and gynaecological surgery sometimes by the use of suppositories.

Toxicity: low—sometimes nausea: metallic taste in mouth.

Lincomycin

Clindamycin (7-chloro-7-deoxylincomycin) is an antibiotic active against Gram-positive bacteria and some anaerobes: previously widely used in hospital because it penetrates tissue well and acts on both *Staph. aureus* and *Bacteroides species*: following a warning notice from the Committee on the Safety of Medicines because it had been associated more often than other drugs with antibiotic-associated pseudomembranous colitis, it is now rarely prescribed: it should not be used if suitable alternative drugs are available.

Erythromycin

The most widely used member of the macrolide group of antibiotics.

Administration: oral, intravenous.

Mode of action: bacteriostatic by inhibition of protein synthesis: may be bactericidal at higher concentrations.

Resistance: lavish use of erythromycin in the 1960s led to emergence of resistant *Staph. aureus*: otherwise this has not been a clinical problem.

Antibacterial spectrum: penicillin-like but includes in addition *H. influenzae, Bord. pertussis, Bacteroides species, Campylobacter species, Legionella pneumophila, M. pneumoniae* and chlamydiae.

Clinical use: staphylococcal infections: a variety of respiratory infections (tonsillitis, sinusitis, bronchitis and pneumonia, diphtheria, whooping cough, atypical pneumonia, psittacosis, Legionnaires' disease): non-specific urethritis: campylobacter enteritis. A useful second-line drug in patients hypersensitive to penicillin.

Toxicity: a safe antibiotic except for one preparation, erythromycin estolate, which may be hepatotoxic.

Vancomycin

Although the drug has been available for almost 30 years, it was little used. Recently, however, indications for its use have become much more common: also the drug is purer and less toxic.

Administration: intravenous, oral.

Mode of action: bactericidal by inhibition of cell wall synthesis: mechanism different from that of β-lactams.

Resistance: virtually unknown.

Antibacterial spectrum: staphylococci (including strains resistant to methicillin and other drugs), streptococci (but less active against *Strep. faecalis*), clostridia.

Clinical use: intravenous—serious infections e.g. endocarditis, septicaemia due to streptococci, coagulase positive and negative staphylococci, especially if multiresistant or if the patient is hypersensitive to penicillins. *Oral*—antibiotic-associated colitis.

Toxicity: when given intravenously phlebitis, ototoxicity, nephrotoxicity: monitor serum levels to control dosage.

OTHER ANTIBIOTICS AND ANTIMICROBIAL DRUGS

Some other less commonly used antibiotics and antimicrobial drugs are listed in Table 45.4.

ANTITUBERCULOUS CHEMOTHERAPY

For more than 20 years, tuberculosis had been successfully treated with a combination of streptomycin, paraamino salicylic acid (PAS) and isoniazid. Problems with streptomycin toxicity, the fact that it had to be given by injection and the unpalatability of PAS caused this regimen to be replaced.

Table 45.4 Some other antimicrobial drugs

Drug	Admini- stration	Mode of action	Antibacterial spectrum	Clinical use	Other features
Fusidic acid	orals i.v.	Bacteriostatic	*Staph. aureus*	Abscesses, osteomyelitis, septicaemia	Good tissue penetration. Resistance may emerge rapidly – give in combination
Chloram- phenicol	oral i.m., i.v., topical	Bacteriostatic: inhibits protein synthesis	Broad- spectrum	Typhoid fever, meningitis, locally eye drops for conjunctivitis	Rarely, causes fatal aplastic anaemia: this has restricted its use
Nalidixic acid*	oral	Inhibits DNA replication	Coliforms, not *Ps. aeruginosa*	Lower urinary infections	Side effects include nausea, rarely visual disturbances
Nitro- furantoin*	oral	Inhibits DNA replication	*Strep. faecalis*, coliforms but not proteus or *Ps. aeruginosa*	Lower urinary infections	Nausea, peripheral neuropathy is sometimes seen
Polymyxins					
E (Colistin)	i.m., i.v.	Bactericidal, lyses bacteria by affecting permeability of cell membrane	*Ps. aeruginosa* many other Gram-negative bacilli but not proteus	Urinary and other infections due to *Ps. aeruginosa* or sensitive coliforms	Toxic effects include nephrotox- icity and neuro- toxicity (para- sthesiae, di..iness)
B	topical			Topical use only for ear, burns, skin infections	
Spectino- mycin	i.m.	Bactericidal	*N. gonorrhoeae* and other Gram-negative bacteria	Penicillin- resistant gonorrhoea	

* Serum levels inadequate for treatment of systemic infections including pyelonephritis

Combination of antimicrobial drugs is essential in tuberculosis to prevent the emergence of resistant bacteria: this was frequent in early days of chemotherapy when single drugs were given.

Antituberculous drugs in current use

1. **Isoniazid** (isonicotinyl hydrazide)
   Administration: oral: (parenteral preparations available for intramuscular, intravenous, intrapleural and intrathecal use).
   Mode of action: bacteriostatic, penetrates well into tissues and fluids and acts on intracellular organisms.
   Resistance: develops readily.
   Toxicity: uncommon, peripheral neuritis, psychotic and epileptic episodes.

2. **Rifampicin**
   Administration: oral.
   Mode of action: inhibits by combining with bacterial DNA-dependent RNA polymerase.
   Resistance: develops rapidly unless other drugs are used in combination.
   Antibacterial spectrum: in addition to mycobacteria, active in vitro against a wide range of Gram-positive and Gram-negative bacteria.
   Clinical use: almost exclusively in tuberculosis but also in prophylaxis of meningococcal meningitis and, occasionally, given in combination with another drug to treat severe staphylococcal infections.
   Toxicity: low, liver function may be affected: often transient hypersensitivity: rarely thrombocytopenia.
   Contraindicated: in the first trimester of pregnancy.
   Patients should be warned that urine, sputum, tears, become coloured red as a result of the therapy.

3. **Ethambutol**
   Administration: oral.
   Resistance: uncommon.
   Toxicity: optic neuritis may develop: it is generally reversible and is uncommon if low dosages are used.

4. **Pyrazinamide**

Other (second-line) drugs which may be used:

Streptomycin
Para-aminosalicylic acid
Thiacetazone

Recommended schedule for pulmonary tuberculosis

Without cavitation: 6 months course of isoniazid, rifampicin ethambutol, pyrazinamide: the latter two stopped after 8 weeks.

With cavitation: treatment should be continued for at least 9 months.

ANTIBIOTIC RESISTANCE IN BACTERIA

A major problem in antibiotic therapy is the emergence of drug-resistant bacteria. The frequency of this depends on the organism and the antibiotic concerned. Some organisms rapidly acquire resistance, e.g. certain coliforms, *Staph. aureus:* others rarely do so, e.g. *Strep. pyogenes.* Resistance to some antibiotics virtually never develops, e.g. metronidazole, whereas with others resistant strains readily emerge, e.g. penicillin, tetracycline, streptomycin.

Drug resistance in clinical practice is associated with antibiotic use: when a small number of resistant bacteria have emerged they will be at a selective advantage in the presence of the antibiotic and will multiply at the expense of sensitive bacteria. The widespread, often indiscriminate prescribing of antibiotics in hospitals has therefore favoured the survival and increase of drug-resistant bacteria.

Drug resistance is of two types:

1. **Primary resistance**
   An innate property of the bacterium and unrelated to contact with the drug, e.g. the resistance of *Esch. coli* to penicillin.

2. **Acquired resistance**
   Due to *mutation* or *gene transfer.*
   Mutation: when resistant bacteria arise as a result of spontaneous mutation: this may be relatively infrequent or—in the case of *Myco. tuberculosis* and streptomycin—common.
   Gene transfer—a major cause of resistance in bacteria—enables resistance to spread from bacterium to bacterium not only within a species but crossing between different species in the case of many coliforms. See Chapter 4.
   Cross resistance. This is when resistance to one antimicrobial drug confers resistance to other—usually chemically related—drugs, e.g. bacteria resistant to one tetracycline or one sulphonamide are resistant to all tetracyclines or all sulphonamides respectively.
   Conversely, *dissociated resistance* is also seen when resistance to one drug is not accompanied by resistance to closely related drugs, e.g. resistance to gentamicin is not always associated with resistance to tobramycin.

Mechanisms of antibiotic resistance

There are three main mechanisms:

1. *Permeability:* the cell membrane becomes altered—possibly by modification of proteins in the outer cell membrane—so that antibiotics or other antimicrobial drugs cannot enter and be taken up by the bacterial cell: often associated with low level resistance to several drugs, but occasionally with high level resistance to a single drug (e.g. tetracycline): a common type of antibiotic resistance in *Ps. aeruginosa*.

2. *Modification of the site of action* of the antimicrobial drug in which the enzyme or substrate with which the drug reacts becomes resistant to it and is able to function normally in the presence of the drug, e.g. trimethoprim and sulphonamide resistance: in the case of trimethoprim resistance, a plasmid or transposon coding for a resistant dihydrofolate reductase is acquired by the bacterium— the bacterium retaining its own chromosomal-coded trimethoprim-sensitive enzyme.

3. *Inactivation of the antibiotic:* is a common mechanism of resistance: the antibiotic is inactivated by enzymes produced by the bacterium; again for the most part plasmid-coded: e.g. β-lactamase destruction of the β-lactam ring responsible for the antibacterial action of penicillins: acetylating, adenylating and phosphorylating enzymes in the case of the aminoglycosides.

PRINCIPLES OF ANTIMICROBIAL THERAPY

1. *Administration* is indicated for an established infection that makes a patient sufficiently ill to require specific treatment: trivial, self-limiting infections in healthy individuals should not be treated with antibiotics.

2. *Choice of drug:* successful chemotherapy depends on the infecting organism being sensitive to the drug chosen. Attempts to treat *viral* respiratory infections—a common practice— are doomed to failure unless there is secondary *bacterial* infection.

   The choice of drug is based on:

   a. *Clinical diagnosis:* implies prescribing on an informed best-guess basis: most infections requiring antibiotics have in fact to be treated before laboratory results are available: they vary in severity from exacerbations of chronic bronchitis to septicaemia. Sometimes the clinical diagnosis indicates a specific bacterial cause, e.g. boil, typhoid fever, but often the cause cannot be

deduced from the clinical picture, e.g. peritonitis, urinary infection.
All medical students must therefore have a working knowledge of the bacteriology of infection so that they can prescribe effectively.

b. *Laboratory diagnosis:* adequate specimens for diagnosis should, whenever possible, be taken before chemotherapy is started. Isolation of the pathogen and sensitivity tests take time—24 h at least and usually longer—and as soon as results are available, treatment should be reviewed. Laboratory monitoring can make the difference between success and failure of therapy.

3. *Route of administration:* drugs must be given parenterally to seriously ill patients. Oral antibiotics for the treatment of systemic infections must be both acid stable (e.g. penicillin V is acid stable but penicillin G is not) and absorbed from the gastrointestinal tract.

4. *Dosage:* must be adequate to produce a concentration of antibiotic at the site of infection greater than that required to inhibit the growth of the infecting organism. *In renal failure,* the dosage of drugs eliminated by the renal route may require either major adjustment (e.g. aminoglycosides, vancomycin) or some more minor modification (e.g. β-lactams) whereas those eliminated by the hepatic route (e.g. erythromycin) can usually be given in normal dosage.

5. *Duration:* treatment of some severe infections e.g. endocarditis, tuberculosis, requires to be prolonged and is aimed at eradication of the pathogen but the majority of acute infections respond to a short course of antibiotics leaving the body defence mechanisms to cope with any infection that remains.

6. *Distribution:* the drug must penetrate to the site of infection, e.g. in meningitis the antibiotic must pass into the CSF. Deep-seated sepsis is a particular problem and an important cause of antibiotic failure: antibiotics cannot penetrate 'walled-off' abscesses or internal collections of pus, and treatment will probably fail unless the pus is drained. Surgical intervention is also necessary if there are established pathological changes, e.g. urinary obstruction due to stones, chronic tuberculous cavities.

7. *Excretion:* agents used to treat urinary infections are excreted in the urine in high concentrations; some, e.g. nalidixic acid and nitrofurantoin, do not achieve useful serum levels but are eliminated almost exclusively by the renal route and are therefore indicated only for the management of lower urinary infections.
Urinary pH affects the activity of some drugs, e.g. the aminoglyco-

sides are far more active in an alkaline medium; the reverse is true of nitrofurantoin, which therefore should not be used to treat infections caused by *Proteus* species which raise the pH of the urine. Erythromycin is excreted largely in the bile; only low concentrations can be detected in urine.

8. *Toxicity:* although the antibiotics in general use are well-tested safe drugs, patients should be warned of possible side effects e.g. the mild diarrhoea that is common with tetracycline therapy or the red colouring of body fluids with rifampicin.

   Serious toxicity can manifest itself in two ways:
   a. *Direct toxicity:* e.g. ototoxicity with the aminoglycosides; nephrotoxicity with vancomycin and the rare bone marrow aplasia due to chloramphenicol.
   b. *Hypersensitivity* is most often due to the penicillins.
      Other complications include *superinfection* with antibiotic-resistant microorganisms, e.g. coliforms and yeasts: this is surprisingly uncommon in immunologically normal patients, e.g. chronic bronchitics on repeated courses of tetracycline, but is a major problem in compromised hosts. Antibiotic-associated pseudomembranous colitis is probably a special example of this.

9. *Use of drugs in combination:* this may be necessary to treat a mixed infection if there is no single agent active against all the causal organisms e.g. in peritonitis due to coliforms and non-sporing anaerobes it is usual to give an aminoglycoside and metronidazole. In addition to this indication, there are two possible advantages:
   a. *Emergence of drug-resistant bacteria will be prevented:* certainly true in the treatment of tuberculosis and may apply in infections due to *Staph. aureus*—an organism with a marked propensity to become resistant—especially when an antibiotic is being given to which resistant strains can emerge during clinical therapy (e.g. fusidic acid).
   b. *Enhanced antibacterial effect (synergism) will be achieved:* this is governed to some extent by the *'Jawetz Law'* on combined action although there are exceptions. It depends on whether each component is bacteriostatic or bactericidal in action and predicts that the effect of a combination will be as follows:

Bactericidal + bactericidal:	may be synergistic
Bactercidal + bacteriostatic:	may be antagonistic
Bacteriostatic + bacteriostatic:	will be additive

   This working rule, however, has some value in clinical practice. The combination of a penicillin and an aminoglycoside, both

Table 45.5 Antibiotic prophylaxis

Clinical situation	Bacteria most likely to cause infection	Appropriate prophylactic regimen
Colonic surgery	Coliforms, anaerobes (bacteroides, anaerobic cocci, clostridia)	Metronidazole plus gentamicin or cephalosporin
Appendicectomy	Anaerobes (as above)	Metronidazole
Gynaecological surgery	Anaerobes (as above)	Metronidazole
Biliary tract surgery	Coliforms	Cephalosporin or cotrimoxazole
Urological surgery if urine infected	Coliforms	As indicated by in vitro tests or gentamicin
Open heart surgery	*Staph. aureus,* *Staph. epidermidis,* corynebacteria	Penicillin plus cloxacillin plus gentamicin
Insertion of prosthetic joints	*Staph. aureus,* *Staph. epidermidis,* corynebacteria	Cloxacillin alone or plus gentamicin
Amputation of ischaemic limb	*Cl. perfringens*	Penicillin
Dental extraction in patients with heart valve disease	Oral streptococci	Amoxycillin or erythromycin
Prevention of tetanus* after wounding	*Cl. tetani*	Penicillin

*In conjunction with immunoprophylaxis.

bactericidal drugs, is often synergistic with an increase in the efficacy of antibacterial action and this may be essential e.g. to treat septicaemia successfully in immunocompromised neutropaenic patients, to eradicate infection in endocarditis. There was disastrous antagonism between penicillin and tetracycline, a bactericidal and a bacteriostatic drug, when these were used in the treatment of pneumococcal meningitis.

10. *Antibiotic prophylaxis:* early but indiscriminate attempts to prevent

infection by giving antibiotics for several days or more failed. This was because the infection to be avoided e.g. pneumonia in unconscious patients or wound infection after surgery, was due to a number of different bacteria not all of which were sensitive to the drugs chosen: prolonged antibiotic administration therefore resulted in the selection of resistant organisms which subsequently caused infection. Successful prophylaxis is possible only when the pathogens are always, or almost always, sensitive to the drug or drugs prescribed.

Short-term prophylaxis: one or at most a few doses of carefully chosen antibiotics given to cover the time when the risk of an infection being established is greatest—i.e. usually the peroperative period. This is a controversial issue. Some examples are listed in Table 45.5.

Short term (2-3 day) prophylaxis should also be given to close contacts of a patient with *meningococcal meningitis,* especially young children. Rifampicin is the drug of choice, minocycline the alternative: sulphonamides are effective if the strain is sensitive.

Long-term prophylaxis: may be continued for months or years.

Rheumatic fever: recurrence of rheumatic fever invariably follows a throat infection with *Strep. pyogenes,* which is always penicillin sensitive. The incidence of further attacks is greatly reduced by giving penicillin.

Urinary tract infection: may be avoided in women who suffer repeated episodes by administration of cotrimoxazole or nitrofurantoin since some 90 per cent of urinary pathogens are sensitive to these drugs.

Tuberculosis: close contacts of a case of open tuberculosis should be given rifampicin and isoniazid for six months.

11. *'Chemotherapy without bacteriology is guesswork':* collaboration between clinician and bacteriologist is crucial especially in the severe infections increasingly encountered in hospitals today. Frequent discussions between ward and laboratory are to be encouraged; they are of benefit to both.

46

Prophylactic immunisation

The introduction of immunisation against infectious disease has been one of the most successful developments in medicine.

Immunisation aims to produce immunity to a disease artificially and without ill-effects.

Immunity can be classified under two main headings:

IMMUNITY

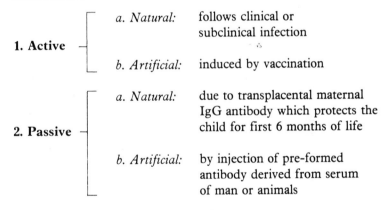

1. Active	*a. Natural:*	follows clinical or subclinical infection
	b. Artificial:	induced by vaccination
2. Passive	*a. Natural:*	due to transplacental maternal IgG antibody which protects the child for first 6 months of life
	b. Artificial:	by injection of pre-formed antibody derived from serum of man or animals

Immunisation can therefore be used to produce both active and passive immunity.

1. Active immunity

Administration of vaccine to provoke an immune response with production of antibody; sometimes cell-mediated immunity is also produced. The onset of immunity is delayed but when established lasts for years, sometimes for life.

   a. *Antibody:* protects in different ways depending on the type of disease; most effective in virus diseases because antibody neutralises virus infectivity. In bacterial disease due to exotoxin, antibody neutralises the toxin: in both bacterial and virus infections antibody enhances phagocytosis.

b. *Cell mediated immunity* (*delayed hypersensitivity*): stimulated independently of antibody: particularly important in resistance to chronic bacterial infections characterised by intracellular parasitism (e.g. tuberculosis, leprosy, brucellosis).

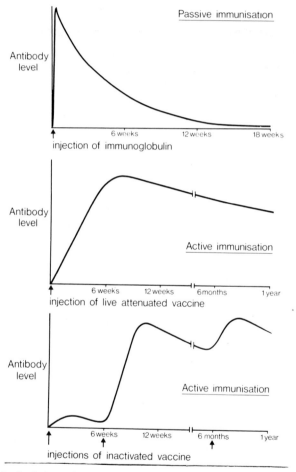

Fig.46.1 Antibody levels following different methods of immunisation.

2. Passive immunity

Injection of pre-formed antibody present in human or animal

serum: immediate immunity is conferred but it is short-lasting—usually for only a matter of weeks.

Figure 46.1 shows antibody levels after passive immunisation and active immunisation with live attenuated and killed organisms.

ACTIVE IMMUNISATION

Objective: to administer vaccine to produce immunity with adequate antibody levels and a population of cells with immunological memory.

Long-lasting: once produced the immunity persists and even after many years infection may still stimulate an accelerated antibody response.

VACCINES

Vaccines are of three types:

1. Live attenuated organisms

Multiply in the body and mimic natural infection with antibody production but without symptoms: reactions are mild and are similar to the natural disease; a single dose gives long-lasting immunity which can be reinforced with later booster doses.

2. Killed (inactivated) organisms

Several—usually three—doses are required because there is no multiplication in the body: generally the first two 6 weeks apart, the second and third 6 months apart; later booster doses are also necessary; reactions do not resemble those of the natural disease and usually follow soon after inoculation.

3. Toxoids

Very successful vaccines in diseases due to a single exotoxin, e.g. diphtheria, tetanus: toxoids are toxins rendered harmless—usually by formaldehyde—but retaining antigenicity. Antigenicity of toxoids can be increased by absorption to a mineral carrier (such as aluminium salts) or by mixture with a suspension of other bacteria containing lipopolysaccharide endotoxin, e.g. the pertussis component of triple vaccine (pertussis, diphtheria, tetanus).

ASSESSING PROTECTION

Immunisation is of value only if it gives a significant degree of protection against infection. Live-attenuated organisms and toxoids are better vaccines than killed organisms. Widespread vaccination of human populations—the aim is usually to achieve acceptance rates of 70 per cent or over—has caused a dramatic fall in the incidence of many infectious diseases, e.g. diphtheria, poliomyelitis: however, in some instances when the introduction of a vaccine has coincided with other measures to control the spread of infection (e.g. improved housing, sanitation, nutrition or the use of chemotherapy which reduced the duration of infectivity) these may have had more effect on the incidence of the disease than vaccination, e.g. tuberculosis.

Field trials, in which the effect of a vaccine in an at-risk population is studied, are of crucial importance: *observe for* a significant reduction in the attack rate or the severity of the disease in vaccinees. Extensive, carefully controlled field trials in human populations are essential in the final evaluation of most vaccines.

Antibody levels (titres) after vaccination and occasionally the appearance of cell-mediated immunity (e.g. tuberculin conversion after BCG vaccination) are estimated routinely when a vaccine is being developed. Detailed studies in human volunteers are easy to organise and the results indicate the likelihood of the vaccine providing worthwhile protection. However, the ability to stimulate antibody production does not guarantee effective prophylaxis. The reasons for the failure of an apparently promising vaccine are complex: in some instances it may be that an inappropriate antibody is formed, e.g. IgM and IgG antibody but not secretory IgA are formed after parenteral administration of killed vaccine although secretory IgA antibody is the principal protective antibody at mucosal surfaces and therefore of great importance in respiratory virus infections (e.g. influenza) and also in poliomyelitis and cholera (where the gut is the site of primary multiplication).

CONTROL OF VACCINE PREPARATION

The manufacture of each batch of vaccine is subjected to stringent quality control to ensure safety and potency.

Safety

The following are the main problems—fortunately now very rare.
1. Contamination: usually of attenuated viral vaccines with other live viruses from the tissue culture used for virus propagation.
2. Inadequate inactivation of killed vaccine.
3. Reversion to virulence of attenuated vaccine.
4. Residual toxicity of toxoids.

Potency

This is ensured by the following measures:
1. Live vaccines are prepared from organisms grown for only a few passages from the parent strain.
2. Killed vaccines are prepared from organisms which have a full complement of the antigens involved in the protective immune response.

Tests of potency measure the antibodies produced in response to inoculation of the vaccine into experimental animals: other tests assess the survival of immunised animals after challenge with virulent organisms but the method of the test may bear little relationship to the human disease e.g. assay of pertussis vaccine tests the protection of mice against intracerebral injection with *Bordetella pertussis* and this is said to correlate with the ability of the vaccine to prevent human disease.

ADMINISTRATION

1. Age

Immunisation aims at the age group at greatest risk: some vaccines (e.g. typhoid and cholera) are indicated regardless of age for anyone entering an endemic area: others (e.g. influenza) are given to those, usually middle-aged or elderly, with chronic cardiac or respiratory·disease.

Childhood: most vaccines, however, are given to children because most of the diseases they prevent are encountered in childhood: e.g. more than two-thirds of the deaths from pertussis are in infants under one year old.

There are two problems in starting immunisation in the first months of life.
a. *The infant immune system is not fully developed* at birth and

capacity to make antibody is therefore limited: nevertheless it is probably adequate if the vaccine is potent.

b. *Transplacental maternal antibody* may prevent a response to live virus vaccines and reduce that to some killed vaccines.

Official policy in Britain is to start immunisation at 3 months old: this is a compromise; delay until the child is 6 to 9 months old would produce better responses but the chance of establishing immunity when it is most needed would be lost.

2. Combined vaccines

Giving more than one vaccine at a time is attractive because it reduces the number of injections required and therefore increases acceptability by the parent and child.

Combined vaccines may enhance antibody production (e.g. the presence of the pertussis component acts as an adjuvant for the toxoids in triple vaccine): but sometimes the response to one organism diminishes that to others; the effect of combining vaccines can only be determined after much trial and research.

3. Live vaccines

These must generally be given at least 3 weeks apart: if not, the risk of reactions can be increased; there may also be interference so that multiplication of one of the viruses is suppressed, preventing an immune response.

COMPLICATIONS OF IMMUNISATION

Side effects: are common after administration of some vaccines, e.g. many killed bacterial vaccines cause local (pain and redness at the injection site) and general (fever and constitutional upset) reactions; although unpleasant these are usually trivial.

Serious complications: although rare, are associated with some vaccines and mainly affect the CNS: they can result in permanent brain damage.

The small but definite risk of serious reactions in a tiny proportion of vaccinees must be balanced against the benefits to the larger number protected: the controversy some years ago over pertussis vaccine illustrates the difficulty that this may present.

Contraindications to vaccination

1. Previous severe local or generalised reaction to that vaccine.
2. Live vaccines should never be given to
   a. *Immunocompromised patients*—because of the risk of severe generalised infections.
   b. *Pregnant women*—because of the danger of transplacental spread to the fetus.

Table 46.1 Vaccines in current use

	Bacterial vaccines	Viral vaccines
Live	BCG (tuberculosis)	Poliomyelitis (Sabin) Measles Mumps Rubella Yellow fever
Killed	Typhoid Cholera Pertussis Pneumococcal extract	Influenza Poliomyelitis (Salk) Hepatitis B Rabies
Toxoids	Diphtheria Tetanus	

VACCINES IN CURRENT USE

Table 46.1 shows a comprehensive list of bacterial vaccines; also included are the main virus vaccines in current use.

LIVE VACCINES

BCG vaccine

Contains: live *Mycobacterium bovis* attenuated by propagation in a bile-potato medium (Bacille Calmette-Guérin). Killed vaccines are of no value as immunising agents in tuberculosis since they do not produce a cell-mediated response.

Indications: policy in the UK is to vaccinate all children between their 10th and 14th birthdays *after* a tuberculin test has shown that they are non-reactors. Give at birth to infants at high risk of contact with tuberculosis, e.g. those known to have a close relative with the disease.

Administration: one dose intradermally over the deltoid area. Normally a red papule develops at the site of injection after some weeks and soon subsides.

Adverse reactions: the papule may progress to an indolent ulcer and discharge pus: associated axillary lymphadenopathy may develop.

Protection: MRC field trials in the UK (1950-71) and studies in North America both showed durable (10-15 year) protection. The incidence of clinical disease in vaccinees was reduced by 80 per cent.

However, other field trials have yielded less encouraging results.

Poliomyelitis vaccine

Contains: live, attenuated strains of poliovirus types 1, 2 and 3—Sabin vaccine. (Salk vaccine which was developed earlier contains inactivated strains of poliovirus types 1, 2 and 3: it is no longer in routine use in Britain.)

Indications: active immunisation of all infants starting when 3 months old.

Administration: orally: three spaced doses are required to ensure multiplication in gut with both local—i.e. gut IgA—and serum antibody production to each of the three types. Booster doses are recommended at school entry and on leaving school.

(Salk vaccine is given by injection and produces serum antibodies only.)

Adverse reactions: minimal: rare cases of paralysis in adults due to Sabin type 3 virus.

Protection: excellent with both vaccines. Following a vigorous campaign with Salk vaccine in the UK launched in 1956 there was a dramatic decrease in poliomyelitis notifications with the vaccine giving an 80 per cent protection against paralytic disease. Sabin vaccine introduced in 1961 has been equally effective and has helped eliminate wild virus from circulation in the community.

Sabin vaccine is preferred to Salk vaccine because:
1. It produces gut immunity.
2. It is easier to administer and can be given more quickly in the face of an epidemic.

However, in some circumstances Salk vaccine is preferred e.g. Third World countries where naturally-occurring enterovirus infection of the gut is common and may interfere with gut multiplication of vaccine virus. Scandinavian countries traditionally prefer Salk vaccine—the newer preparations of which are considerably more potent than the earlier vaccine.

Measles vaccine

Contains: live attenuated virus.

Indications: active immunisation of all children in the second year of life to prevent the respiratory complication of measles and the mental retardation that may follow measles encephalitis.

Although the disease is usually mild in the UK, in some developing countries measles is very severe—mortality rates around 10 per cent have been recorded in tropical Africa. The acceptance rate of vaccination in the UK is now only about 50 per cent.

Administration: one dose by injection.

Adverse reactions: few, the present vaccine is well-tolerated: fever and transient rash—like mild measles—may follow 6-12 days after vaccination. Post infectious encephalomyelitis, a serious complication, is rare—about 1 case per million.

Protection: good and apparently long-lasting; probably slightly less solid than that following natural infection; so far there is no evidence that immunity deteriorates significantly with increasing age. It would be serious if childhood immunity were to wane and leave some people susceptible in adult life.

Mumps vaccine

Although available, is used very little in Britain and is not one of the officially recommended vaccines.

Contains: live attenuated virus.

Indications: prevention of mumps in children over one-year old and in adults.

Administration: one dose by injection.

Adverse reactions: few; occasionally fever, very rarely parotitis.

Protection: substantial durable immunity for 10 years at least, probably for much longer.

Note: Mumps vaccine is widely used in the USA alone or in combination with measles and/or rubella vaccine.

Rubella vaccine

Contains: live attenuated virus.

Indications: in the UK the vaccine is administered to all girls between 10 and 14 years of age and to susceptible adult women post-partum. This policy, aimed at preventing congenital rubella, does not affect the epidemiology of the disease and outbreaks of rubella have continued to take place (in which some susceptible pregnant women have contracted infection).

Note: A past history of rubella not confirmed by laboratory tests is an unreliable guide to immune status: other virus diseases can mimic rubella clinically.

Administration: one dose by injection.

Adverse reactions: uncommon but there may be mild rubella-like symptoms including arthralgia some 9 days after vaccination. Pregnancy must be avoided for 3 months after vaccination: the vaccine must never be given during pregnancy.

Protection: apparently good with long-lasting immunity.

KILLED VACCINES

Typhoid vaccine

Contains: heat-killed phenol-preserved suspension of *Salmonella typhi.*

Indications: for those travelling to or living in areas where typhoid fever is endemic.

Administration: two doses 4 to 6 weeks apart by injection: booster doses every 3 years.

Adverse reactions: local and general reactions are common: early in onset, they subside in 36 hours.

Protection: around 70 to 90 per cent: extensive field trials in Yugoslavia (1954-55) carried out by WHO showed a good protection rate against typhoid. An alcohol-killed, alcohol-preserved vaccine (alcohol preserves the Vi—'virulence'—antigen) evaluated at the same time was ineffective.

Cholera vaccine

Contains: heat-killed *Vibrio cholerae* serotypes Inaba and Ogawa.

Indications: for those travelling to endemic areas or countries which require evidence of vaccination.

Administration: two spaced doses by injection; booster doses every 6 months.

Adverse reactions: local and general reactions are quite common; serious reactions are rare.

Protection: poor, estimated at 50 per cent, and short-lasting —about 6 months: there is a similar degree of cross-protection to the El Tor biotype which is antigenically identical.

Attempts have been made to produce better vaccines, e.g. live avirulent oral vaccines to stimulate secretory IgA antibodies in the gut (coproantibodies) as well as serum antibodies and a toxoid prepared from cholera enterotoxin (exotoxin): none has so far gained acceptance.

Pertussis vaccine

Contains: killed, freshly-isolated, smooth strains of *Bordetella pertussis*. The vaccine should contain all the surface antigens of *Bord. pertussis* associated with epidemics: these antigens designate the three common serotypes—1,3; 1,2,3; and 1,2. However, it is still not certain if these antigens are responsible for a protective response. Manufacturers select their own strains for vaccine production and there are significant differences in the protection afforded by individual vaccines and in the incidence of adverse reactions they cause.

Indications: official policy in Britain is the active immunisation of all children starting at 3 months old. The heated debate which began in 1974 about the risk and efficacy of vaccination aroused much public concern. As a result, the acceptance rate declined and major outbreaks of pertussis resulted in 1977/79 and 1981/82.

Administration: three doses at intervals of 6 to 8 weeks and, after, the second, 4 to 6 months: booster doses are not recommended because pertussis is not a problem after 5 years of age.

Adverse reactions: usually local and trivial: one infant in 40 develops excessive crying and irritability. Severe reactions attributed to the vaccine are convulsions, severe neurological reaction (1 case per 100 000 injections) and permanent brain damage (1 case per 300 000 injections).

Protection: Mass vaccination was started in the UK in 1957. During the following years there have been conflicting claims about the efficacy of the vaccine. The protection is not solid but is of the order of 80 per cent. Whooping cough is a disease that has become less severe as living conditions have improved and children have become healthier: but it has not been eliminated. Nevertheless, the case fatality rate has fallen dramatically and there is good evidence that vaccination has prevented epidemics.

Pneumococcal vaccine

Contains: a saline solution of 23 highly purified *capsular polysaccharides* extracted from pneumococci of the most prevalent pathogenic types, i.e. 1,2,3,4,5,6B,7F,8,9N,9V,10A,11A,12F, 14,15B,17F,18C,19A,19F,20,22F,23F and 33F. Not generally available in U.K.

Indications: to prevent pneumococcal pneumonia, bacteraemia and meningitis in individuals at special risk.

Administration: one dose by injection.

Adverse reactions: local in about half and fever in 10 per cent of those vaccinated.

Protection: apparently good—but only against infections caused by the serotypes present in the vaccine.

Influenza vaccine

Contains: inactivated virus, usually two of the currently circulating strains of influenza A virus with the current influenza B strain also included. Different vaccines contain whole virus, 'split' virus (i.e. partially purified disrupted particles) or surface antigens (highly purified haemagglutinin and neuraminidase).

Indications:

1. Elderly people with pre-existing cardio-respiratory or renal disease, especially useful for old people in residential homes or long-stay hospitals
2. Key personnel in essential services, e.g. hospital staff, the police force.
   Administration: one dose by injection. Vaccination requires to be repeated each winter.
   Adverse reactions: few and mild but a number of cases of polyneuritis (Guillain-Barré syndrome), some severe, were recorded in USA following mass vaccination in 1976-77; there may be severe reactions in those hypersensitive to egg protein.
   Protection: short-lived,—i.e. about a year, the protection conferred is of the order of 70 per cent.

A live-attenuated vaccine administered intranasally has had trials in Eastern Europe; it stimulates the production of secretory IgA on the respiratory mucous membranes but causes more side effects and has not gained acceptance.

TOXOIDS

Diphtheria toxoid

Contains: diphtheria 'formol toxoid' (i.e. toxin treated with formaldehyde)—available also absorbed onto aluminium phosphate or hydroxide.

Indications: children and selected 'at-risk' adults, e.g. hospital or laboratory staff.

Administration: three spaced injections starting at 3 months old as for pertussis vaccine with which it is usually combined as part of the Triple Vaccine. Booster dose at school entry.

Older children (i.e. aged 10 years or more) and adults: use a low dose (1.5 Lf or flocculating units) vaccine: administer by deep subcutaneous or intramuscular injection.

Adverse reactions: mild and transient under 10 years of age: older children and adults may experience severe side effects and in this group a preliminary Schick test is advised: toxoid is given only to those who are positive reactors; pseudo reactors should not be immunised.

Protection: excellent—the disappearance of diphtheria in the UK between 1941 and 1951 was due to immunisation and the disease is now extremely rare in this country. However, every few years a small outbreak—usually in the unvaccinated—is reported which can be missed by the unwary or unprepared.

Tetanus toxoid

Contains: tetanus formol toxoid—available also absorbed onto aluminium hydroxide.

Indications: the aim is active immunisation of the entire population; although the disease is rare, tetanus may develop after common, trivial wounds.

Administration: three spaced injections starting in infancy as part of the Triple Vaccine. In the unvaccinated, a course should begin when a situation of risk presents, e.g. after injury at the casualty department. Booster doses at 5 years and another 5 to 15 years later and in the event of injury.

Adverse reactions: rare and minor; severe reactions restricted to adults, usually those who have been hyperimmunised—i.e. who have had too many booster injections.

Protection: excellent.

Triple vaccine

Contains: killed *Bordetella pertussis,* diphtheria toxoid and tetanus toxoid.

Indications: active immunisation of all infants.

Administration: three spaced doses by injection. At school entry, booster doses of diphtheria and tetanus toxoids only.

Adverse reactions: see individual vaccines.

Protection: see individual vaccines.

The scheme for active immunisation officially recommended in the United Kingdom is shown in Table 46.2.

Hepatitis B vaccine

Contains: inactivated hepatitis B surface antigen (HBsAg) purified and absorbed in aluminium salt.

Indications: those at special risk e.g. health care personnel in mental deficiency hospitals, renal units, laboratory workers, those handling blood products. Infants not less than 6 months old, born to HBsAg-carrier mothers.

Administration: in three doses intramuscularly separated by 1 and 6 months respectively.

Adverse reactions: local pain and redness.

PASSIVE IMMUNISATION

Objective: to produce immunity immediately by the injection of antibodies present in human or animal serum. These antisera are also used in treatment.

Short-lasting: the immunity that follows wanes in a matter of weeks or a few months.

Table 46.2 shows the current recommended schedules of immunization in Britain.

SPECIFIC IMMUNOGLOBULINS

1. Human:

Prepared from plasma pools containing high levels of the appropriate antibody. Donations are taken from individuals after their plasma has been screened for antibody content: sometimes they are chosen because they are recovering from infection (convalescent serum) or have recently been actively immunised against the disease. Human immunoglobulins have a half life of 26 days after

Table 46.2   Schedule of vaccination and immunisation recommended in the United Kingdom

Age	Vaccine	Notes
During the first year of life	Triple Vaccine and polio vaccine. Triple Vaccine and polio vaccine Triple Vaccine and polio vaccine	6-8 weeks between the 1st and 2nd doses; 4-6 months between the 2nd and 3rd doses. First dose should be given 3 months
During the second year of life	Measles vaccine	
5 years (school entry)	Diptheria-tetanus toxoids Polio vaccine	Booster dose Booster dose
Between 10th and 14th birthdays	BCG vaccine	For tuberculin-negative children
Between 10th and 14th birthdays	Rubella vaccine	Administer to *all girls* irrespective of past history of an attack of rubella
On leaving school	Polio vaccine Tetanus toxoid	Booster dose Booster dose
Adult	Polio vaccine (if previously unvaccinated)	For travel to tropical countries or parents of child being given polio vaccine
Adult woman of childbearing age	Rubella vaccine	For seronegative women at occupational risk (teachers, nurses) or after pregnancy
Adult	Tetanus toxoid	For previously unvaccinated adults especially following a contaminated wound.

Note: live vaccines must be administered not less than three weeks apart.

injection and significant protection may last up to 3 months and sometimes longer.

Preparations

Some of the available preparations are listed below:

Hepatitis B immunoglobulin: indication: post-exposure prophylaxis after accidental inoculation with HBsAg positive blood.

Tetanus immunoglobulin: indication: prophylaxis and treatment of tetanus.

Varicella-zoster immunoglobulin: indication: treatment of the disease in immunocompromised hosts who may develop severe varicella or generalised zoster.

Diphtheria immunoglobulin: indication: prophylaxis and treatment of diphtheria.

2. Animal:

Specific antibodies, raised in horses by active immunisation, were used extensively in the past. Unfortunately they contain foreign protein to which the recipient forms antibody with the result that:

1. They are rapidly eliminated—much faster than human immunoglobulins.
2. They can cause unpleasant and sometimes dangerous hypersensitivity reactions including serum sickness (an Arthus reaction) and anaphylaxis.

Equine antisera still available include diphtheria and botulinum antitoxin and polyvalent gas-gangrene antitoxin (of doubtful value as a therapeutic agent).

NON-SPECIFIC IMMUNOGLOBULINS

Human immunoglobulin prepared from donations of pooled normal plasma: contains antibodies to a wide range of infective agents likely to have been encountered by most people.

Indications:

1. *Prophylaxis of hepatitis A.* Administer to travellers going to endemic areas: family contacts of a case.
2. *Prophylaxis of measles:* given within 6 days of exposure will prevent or modify the disease.
3. *To boost immunoglobulin levels* in children with hypogammaglobulinaemia.

Medical mycology

47

Fungal infections

Fungi, unlike bacteria, are *eukaryotic:* the cell nucleus contains multiple chromosomes enclosed by a membrane and in the cytoplasm there are mitochondria and 80s ribosomes (Chapter 2). Many can reproduce sexually—this process involves meiosis. Most fungi grow as filaments (*hyphae*) which intertwine to form a mesh (the *mycelium*): most yeasts, which reproduce by budding, are exceptions. Of the thousands of species only a few are pathogenic for man: some others cause disease in animals, fish, insects and plants.

Habitat: fungi, like bacteria, are ubiquitous. In the soil they play an important role in the degradation of organic compounds and they may produce antibiotics (e.g. penicillin) which inhibit the growth of competitive bacteria.

Culture: all fungi are aerobic and most grow readily on simple media.

Classification: this is complex and based on the method of spore production (sexual and asexual), the morphology of the colony, the vegetative hyphae which form the mycelium and the specialised aerial hyphae which bear the spores. Fungi of medical importance can conveniently be divided into three groups:
1. Yeasts
2. Filamentous fungi
3. Dimorphic fungi

DISEASES

1. *Infections (mycoses):*
a. *Superficial infections* of the mucosa with yeasts (causing thrush) and the keratin of skin, nail and hair with filamentous fungi called dermatophytes (causing ringworm) are common in the UK. These infections although troublesome are usually trivial and do not involve deeper tissues.

361

b. *Subcutaneous infections,* the result of the traumatic implantation of environmental fungi leading to progressive local disease with considerable tissue destruction and sinus formation, are rare in the UK but common in the tropics.

c. *Systemic infections* with haematogenous spread throughout the body are serious and often fatal. They are uncommon in the UK except in compromised patients with impaired host defences (Ch. 42) who may develop widespread disease due to yeasts or filamentous fungi such as *Aspergillus species.* In other parts of the world, certain forms of deep disseminated mycoses caused by dimorphic fungi are remarkably common in otherwise healthy individuals.

2. *Mycotoxicoses:* the result of eating mouldy food in which the fungus has produced toxic metabolites. Examples are poisoning following the consumption of food containing aflatoxins formed by the growth of *Aspergillus flavus:* ergotism after the ingestion of wheat infected with *Claviceps purpurea.*

3. *Allergic reactions:* inhalation of fungal spores, notably those of *Aspergillus fumigatus,* may provoke a type I and/or a type III hypersensitivity reaction. Sometimes the antigenic stimulus is prolonged because the fungal hyphae grow in the lumen of the bronchi: invasion of lung tissue does not take place.

YEASTS

Yeasts are round to oval unicellular fungi which reproduce by budding. Some may develop pseudohyphae—chains of elongated budding cells—but only a few are able to form true hyphae.

CANDIDA

Several species are found in man but one species, *Candida albicans,* is responsible for 90 per cent of infections. The other species include *C. stellatoidea, C. tropicalis, C. krusei, C. guilliermondii* and *C. parapsilosis.*

Candida albicans

Habitat: the normal flora of the upper respiratory, gastrointestinal and female genital tracts.

Laboratory characteristics

Morphology and staining: two forms, both Gram-positive, are recognized in clinical material and on culture.

1. Spherical to oval budding cells *(3-5×5-10 μm)*. The yeast or Y-form.
2. Elongated filamentous cells joined end-to-end (pseudohyphae) and producing buds (blastospores): also true hyphae. These constitute the mycelial or M-form. *C. albicans* is the only species to produce abundant pseudohyphae *in vivo* (Plate 15).

Culture: aerobic and easy to cultivate but isolation from clinical material may be impeded by faster growing bacteria.

1. Sabouraud's medium: a simple glucose peptone agar, pH 5.6, often made more selective by the addition of antibiotics (e.g. chloramphenicol) is useful for primary isolation. Incubation at 37 °C for 48 h may be necessary.
2. Ordinary agar and blood agar: colonies may be observed more easily around antibiotic discs which have inhibited bacterial growth.

Colonial morphology: Colonies are cream to white in colour, flat or hemispherical in shape and have a waxy surface. The yeasts are predominantly in the Y-form but M-forms develop in older cultures and the pseudohyphae may project from the edge of the colonies.

Identification: can be readily differentiated from other species by production of:

1. *Germ-tubes*
   Method: grow in serum for 3 h at 37 °C, make a wet film and examine for formation of filamentous outgrowths—germ tubes.
2. *Chlamydospores*
   Method: grow in a nutritionally poor medium (e.g. cornmeal agar) for 24 h at 28 °C and examine for presence of round thick-walled resting structures—chlamydospores—usually found at the ends of pseudohyphae deep in the agar.

Biochemical activity: the results of fermentation (anaerobic metabolism) and assimilation (aerobic metabolism) of a range of carbohydrates are used in the identification of *Candida* species.

Antigenic structure: strains fall into two serotypes: A—antigenically similar to *C. tropicalis;* B—antigenically similar to *C. stellatoidea.* Candida antibodies can be demonstrated in most human sera. Delayed-type hypersensitivity is common and a positive candida skin test is almost universal in normal adults.

Pathogenicity

Source: usually endogenous but cross-infection may occur, e.g. from mother to baby, from baby to baby in a nursery.

Host: infections are most common in babies who are premature and adults debilitated by general ill-health, notably diabetes. A special 'at-risk' group is composed of patients compromised either by the nature of their disease (e.g. malignancy, in particular leukaemias or lymphomas) or the treatment they have received (e.g. long courses of wide-spectrum antibiotics, immunosuppressive or cytotoxic drugs).

Lesions: infections (*candidosis*) are usually superficial (mucous membranes and skin) but occasionally are deep and involve internal organs.

Superficial

Mucous membranes: thrush—white adherent patches on buccal mucosa or vagina.

Skin: red weeping areas usually when skin is moist and traumatized, e.g. intertrigo in the obese.

Chronic mucocutaneous candidosis: an intractable disfiguring condition especially affecting the face and scalp: onset generally in infancy. Due to a defective immunological response in which T-cell function is reduced but antibody response to candida remains normal.

Deep

Involvement of lower respiratory tract and urinary tract; septicaemia with localization in endocardium, meninges, kidney, bone.

Diagnosis

1. By demonstration of yeasts in a wet film or Gram-stained smear followed by isolation of candida when the specimen is cultured.
2. By detection of antigen in serum or other body fluid.
3. By detection of antibody in serum. A variety of methods e.g. ELISA are available but the tests are of limited value because antibody is present in more than half of the healthy adult population.
4. By detection using gas liquid chromatography of elevated levels

of arabinitol, a metabolite of candida, in serum. A positive result may indicate systemic infection.

Treatment

Candida are eukaryotic microorganisms and are resistant to *all* antibacterial antibiotics.

Superficial infections can be treated topically with a polyene (nystatin, amphotericin B) or an imidazole (miconazole, clotrimazole). Systemic infections require intravenous amphotericin B either given alone or with 5-fluorocytosine: the combination may be synergistic.

CRYPTOCOCCUS

The one pathogenic species in the genus is *Cryptococcus neoformans*.

Cryptococcus neoformans

Habitat: Ubiquitous saprophyte: often found in soil especially in ground contaminated with bird droppings, particularly associated with pigeons.

Laboratory characteristics

Morphology and staining: a capsulated budding yeast with spherical cells 5–15 μm in diameter. Does not usually form a pseudomycelium. Gram-positive although the capsule may prevent staining.

Culture: aerobic: grows on a wide variety of common media at 37 °C but also at room temperature. Isolation from clinical material is best achieved by culture at 30 °C: several days incubation may be required before colonies develop.

Colonial morphology: on Sabouraud's medium forms glistening mucoid cream colonies which become duller and darker on extended incubation.

Identification

1. Make a wet preparation of a portion of the colony in India ink to demonstrate the capsule. *Note* the capsule which is usually

pronounced in clinical material may be rudimentary in culture.
2. Inoculate mice intraperitoneally: *observe* gelatinous lesions, more marked in the brain, when they die in 2 to 3 weeks.

Biochemical activity: produces a urease: assimilates a number of compounds including inositol but is unable to metabolise by fermentation: produces phenol oxidase detected by the formation of brown colonies on birdseed agar.

Pathogenicity

Pathogenic for man and a variety of animals.

Source: from the environment usually by inhalation especially of dust containing pigeon excreta.

Lesions: a lung granuloma, usually symptomless, is the primary lesion. This resolves spontaneously in the vast majority of patients without dissemination. Haematogenous spread results in subacute or chronic meningoencephalitis—the classic disease presentation—and sometimes involvement of skin, lungs, lymph nodes and other organs. Clinical disease is usually found in immunocompromised patients and may result from the reactivation of an old healed focus.

Diagnosis

1. By demonstration in CSF, exudate, urine or other appropriate specimen of an encapsulated yeast confirmed by isolation on culture. *Note:* CSF changes resemble those in tuberculous meningitis and the yeast may be confused with RBCs or lymphocytes.
2. By detection, using a latex agglutination test, of antigen in CSF, blood or urine.
3. By detection of antibody in serum: a variety of tests are available including agglutination and immunofluorescence.

Treatment

Intravenous amphotericin B combined with 5-fluorocytosine.

MALASSEZIA

Malassezia furfur (formerly known as *Pityrosporum orbiculare* or *P. ovale*)
   Habitat: skin.

Laboratory characteristics

Morphology: oval yeast reproducing by unipolar budding.
Culture: dull, buff colonies on agar supplemented with lipids e.g. olive oil.

Pathogenicity

Cause of tinea versicolor in which large scaling patches develop on the skin of the trunk; brownish on light-skinned people, lighter on dark-skinned people; usually asymptomatic.

Diagnosis

Although the yeast can be isolated on culture, diagnosis is usually made by the demonstration of short curved non-branching hyphae and yeasts in skin scales.

OTHER YEASTS

Torulopsis glabrata (Candida glabrata) is a skin commensal and may be isolated from urine and blood as a contaminant. It can be a cause of infection in immunocompromised patients.
Rhodotorula species are yeasts which form pink colonies. They are skin commensals and may be cultured from contaminated clinical specimens. Grow at room temperature but not 37°C.

FILAMENTOUS FUNGI

DERMATOPHYTES

The dermatophytes are a group of fungi which cause infection— *tinea* or *ringworm*—of skin, nail and hair without involvement of living tissue. They belong to three different genera, *Trichophyton, Microsporum* and *Epidermophyton.* The common species and the types of ringworm they cause are listed in Table 47.1.
   Habitat: The keratin of humans and animals.
   Infection is acquired either by person to person spread, sometimes via fomites or from contact with animals as a zoonosis. A few are present in the soil which acts as the reservoir of infection (see Table 47.1).

Table 47.1 Dermatophytes and ringworm

FUNGUS	SOURCE	TYPE OF RINGWORM CAUSED				
		Tinea capitis	Tinea corporis	Tinea cruris	Tinea pedis	Tinea unguium
Trichophyton rubrum	Human		+	++	+++	+++
Trichophyton mentagrophytes var interdigitale	Human			+	++	+
Trichophyton mentagrophytes var mentagrophytes	Animal — cattle horses, rodents	+	++			
Trichophyton schoenleinii	Human	+ usual cause of favus				
Microsporum audouinii	Human	+	+			
Microsporum canis	Animal — cats, dogs	+++	++			
Microsporum gypseum	Soil	+	+			
Epidermophyton floccosum	Human			++	+	+

+++ = most commonly caused by   ++ = commonly caused by   + = sometimes caused by

Clinical features

These are summarised in Table 47.2.

Table 47.2 Clinical manifestations of ringworm

Site	Affect	Clinical Features
Tinea capitis	scalp and hair	small scaling papules which spread to leave areas of baldness: infected hairs break to leave stumps. Skin may suppurate. Favus is a variety characterised by yellow lesions which later develop cup shaped crusts that heal leaving atrophic bald skin.
Tinea corporis	skin excluding scalp, bearded areas and feet	circular spreading lesions: as the centre scales and heals the periphery advances with vesicles and pustules in inflamed skin.
Tinea cruris	skin of groin and perineum	spreading scaly dermatitis with vesicopustular edge: little central healing
Tinea pedis	soles of feet and between toes	inflamed skin with vesicles leading to peeling and fissure formation (athlete's foot)
Tinea unguium	nails of hands or feet	affected nails appear opaque, thickened, brittle and distorted: they may separate from nail bed and be totally destroyed

Diagnosis

Specimen: scrapings of skin or nail, short lengths of plucked hair.

Examine by:

Direct microscopy: a positive finding establishes the diagnosis of ringworm.

Make a wet preparation of the specimen in 20 per cent potassium hydroxide: leave for 10–20 min to digest keratin.

Observe. Filamentous branching hyphae: spores may also be seen. In tinea capitis, fungal elements may be seen either outside the hair—ectothrix infections e.g. with *Microsporum canis* or, less commonly, inside the hair—endothrix infections e.g. with *Trichophyton schoenleinii*.

Culture: necessary to identify the causal fungus.

Inoculate the specimen onto a plate of Sabouraud's agar containing chloramphenicol—this antibiotic suppresses the growth of contaminating bacteria.

Table 47.3   Identification of dermatophytes

Fungus	Colonial appearance*	Characteristic microscopic features of colony
Trichophyton rubrum	**Surface:** white, granular or fluffy **Reverse:** dark red to brown	**Microconidia:** tear shaped
Trichophyton mentagrophytes var interdigitale	**Surface:** white, woolly **Reverse:** yellow	**Microconidia:** spherical **Hyphae:** spiral
Trichophyton mentagrophytes var mentagrophytes	**Surface:** cream, granular **Reverse:** brown	**As above:** microconidia more abundant
Trichophyton schoenleinii	**Surface:** yellow to brown, waxy, folded **Reverse:** colourless to yellow	**Hyphae:** antler like

Microsporum audouinii

Surface: grey to cream, flat, velvety with radial feather-like markings
Reverse: pink to cream

Hyphae: comb like with terminal chlamydospores

Microsporum canis

Surface: white to yellow, flat, velvety with radial feather-like markings
Reverse: deep yellow

Macroconidia: thick walled, pointed with superficial projections

Microsporum gypseum

Surface: cream, powdery
Reverse: yellow to tan

Macroconidia: thick walled, oval with superficial projections

Epidermophyton floccosum

Surface: greenish brown, suede-like, folded
Reverse: yellow to brown

Macroconidia: club-shaped smooth surface

*After 7–10 days culture on Sabourauds agar at 28°C

Incubate aerobically at 25° to 30°C.

Examine daily up to 21 days for fungal colonies.

Identify the fungus by:

1. *Colonial appearance*
   Observe pigmentation and texture of the surface of the colony and the pigmentation of the reverse side seen through the bottom of the plate (Table 47.3).

2. *Microscopic features of colony*
   Transfer growth to slide carefully so that structural arrangement is preserved: suspend in a drop of alcohol.
   Make a wet preparation by staining growth with lactophenol cotton blue.
   Observe hyphae and conidia. *Conidia* are asexual spores: two types are formed, small unicellular microconidia and larger septate macroconidia. Microscopic morphology aids identification: some characteristic features are listed in Table 47.3.

Treatment

Mild infections: topical imidazole (e.g. clotrimazole or miconazole): this treatment has superseded local applications of benzoic acid (Whitfield's lotion), undecenoic acid, etc.

Severe infections: oral griseofulvin for 4 to 6 weeks, if hair is involved for 3 to 6 months, if nails are involved continue for one year. Oral imidazole (e.g. ketoconazole) is the alternative treatment.

OTHER FILAMENTOUS FUNGI

ASPERGILLUS

Aspergillus fumigatus is the main pathogen: other species associated with infection include *A.niger* and *A.flavus*.

Habitat: soil and dust: spores are ubiquitous.

Laboratory characteristics

Culture: after 3 to 4 days incubation on Sabouraud's agar at 25° to 37°C, the colonies have a velvety to powdery surface and are characteristically coloured: *A.fumigatus* dark green, *A.niger* black on white and *A.flavus* yellow-green.

Microscopic appearance of the colony: a wet preparation stained with lactophenol cotton blue demonstrates septate hyphae and conidiophores—specialised aerial hyphae that bear conidia (i.e. spores). The conidiophores have swollen rounded ends and the spores are formed in chains. The general morphology is characteristic of the genus and there are also inter-species differences that are useful in identification (Fig. 47.1)

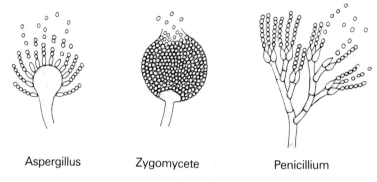

Aspergillus   Zygomycete   Penicillium

Fig. 47.1 Spore bearing structures of some fungi.

Pathogenicity

Aspergillus species can cause a variety of clinical syndromes.
1. *Allergic bronchopulmonary aspergillosis:* inhaled spores provoke a hypersensitivity reaction which may be of
   (a) Type I (asthma)
   (b) Type III (extrinsic alveolitis).
   (c) Both combined i.e. type I and type III
2. *Aspergilloma* in which a fungal ball grows within, and is usually restricted to, an existing lung cavity e.g. due to old tuberculosis, bronchiectasis.
3. *Invasive aspergillosis* in which the fungus establishes a pneumonia and later disseminates to involve other organs, e.g. brain, kidneys, heart. Patients who develop this type of disease are usually immunocompromised (Ch. 42).
4. *Superficial infections* of the external ear (otomycosis) and, less commonly, the eye (mycotic keratitis) and nasal sinuses.

Diagnosis

1. *Direct microscopy* to demonstrate septate hyphae: suggestive, but not diagnostic of, aspergillus infection.

Specimen:

a. Exudate e.g. sputum: make wet preparation in 20 per cent potassium hydroxide.

b. Tissue, e.g. biopsy or post-mortem material: stain sections by PAS method—hyphae are poorly stained by haematoxylin and eosin.

2. *Isolation* by culture on Sabouraud's agar at 25°-37°C. Colonies grow after 48 h but longer incubation may be required before characteristic morphological features develop.

   Note: since aspergillus spores are ubiquitous,· colonies of the fungus are often found growing on cultures as the result of aerial contamination. Thus it may be difficult to interpret the significance of isolating a few colonies from a clinical specimen.

3. *Serology:* precipitating antibodies to aspergillus antigens can be demonstrated by a number of laboratory methods including counter current immuno-electrophoresis, immunodiffusion and ELISA. Antibodies are usually absent in the sera of healthy individuals: they can be detected in the majority (70 per cent) of patients with allergic aspergillosis and approximately the same proportion of those with pneumonia or invasive disease.

Treatment

Invasive aspergillosis is treated with intravenous amphotericin B. The mortality is high.

ZYGOMYCETES

The genera associated with human infection are mucor, absidia and rhizopus.

Habitat: ubiquitous in the soil: spores in air and dust.

Laboratory characteristics

All three genera are similar.

Culture: after 3 to 4 days incubation on Sabouraud's agar at 30° to 37°C, the colonies are grey-white with a thick cottony, fluffy surface.

Microscopic appearance of the colony: non-septate broad hyphae with aerial sporangiophores which end in a sporangium—a sac containing spores (Fig. 47.1).

Pathogenicity

Zygomycosis (mucormycosis, phycomycosis) occurs as a systemic infection following dissemination from a primary focus, often in lung: almost all patients are immunocompromised (Ch. 42). The rhinocerebral form, in which the nose, nasal sinuses and orbit are involved, is a well-recognised and usually fatal complication of diabetes: infection may penetrate to involve the frontal lobe of the brain.

Diagnosis

1. *Direct microscopy* to demonstrate broad non-septate hyphae.
   Specimen:
   a. Exudate: make wet preparation in 20 per cent potassium hydroxide.
   b. Tissue: hyphae stain readily with haematoxylin and eosin (unlike aspergillus).
2. *Isolation* by culture on Sabouradud's agar: may be difficult to achieve from necrotic material even when abundant hyphae are seen.
   Note: these common environmental moulds are frequent contaminants of culture plates.

Treatment

Intravenous amphotericin B combined, where appropriate, with surgical drainage. Good medical control of diabetes.

PENICILLIUM

A variety of species abound in the environment and grow on bread, jam, fruit, cheese, etc. In the laboratory they are common air-borne contaminants of culture media.

Colonies are blue-green in colour with a white border and have a powdery surface.

Microscopy demonstrates septate hyphae with branched conidiophores bearing chains of spores—the appearance likened to a 'brush or broom'. (Fig. 47.1).

FUNGI CAUSING MYCETOMA

Mycetoma usually affects the foot ('Madura foot') and can be caused by a variety of fungi and actinomycetes (usually *Nocardia species*—see p. 109). Important filamentous fungi include *Madurella mycetomi*, *Madurella grisea* and *Phialophora jeanselmei*.

Habitat: the soil.

Pathogenicity

Fungi implanted into subcutaneous tissue following trauma (e.g. by a splinter) produce destructive granulomatous lesions which drain through multiple sinus tracts. There is local spread but no dissemination. A common condition in tropical and subtropical areas where people go barefoot.

Treatment is surgical: chemotherapy ineffective when mycetoma is due to filamentous fungi.

DIMORPHIC FUNGI

Dimorphic fungi grow either as yeasts or as filaments. *Yeast form* (parasitic phase) is found in infected tissues and on artificial media at 37 °C. *Filamentous form* (saprophytic phase) is present in the soil and on artificial media at 22° to 25 °C.

Habitat: the soil: some have a characteristic geographical distribution (Table 47.4).

Table 47.4 Dimorphic fungi and disease

Fungus	Disease	Geographical distribution
Blastomyces dermatidis	North American blastomycosis	North America especially Mississippi and Ohio valleys
Paracoccidioides brasiliensis	South American blastomycosis	South America: less commonly, Central America
Coccidioides immitis	coccidioidomycosis	USA from California to Texas: South and Central America
Histoplasma capsulatum	histoplasmosis	Eastern and Central USA: occasionally other parts of the world
Histoplasma duboisii	African histoplasmosis	equatorial Africa
Sporothrix schenckii	sporotrichosis	worldwide

Pathogenicity

Cause disease in man (Table 47.4), wild and domestic animals.

Infection is usually acquired by inhalation and the primary lesions are in the lungs. In most cases these heal often without causing illness and delayed hypersensitivity, with a positive skin test reaction to the appropriate antigen, develops. Progressive disease may affect the lungs, sometimes causing cavitation, and/or disseminate widely to involve the skin, mucous membranes and internal organs. The lesions are chronic granulomas.

Note: the similarity of this disease process to tuberculosis.

Sporotrichosis is different: it follows traumatic implantation of the fungus into the skin and results in a chronic local pyogenic infection with lymphatic spread and ulceration of the lymph nodes: disseminated disease is rare.

Diagnosis

1. *Direct demonstration:* of the yeast-like form in suitably stained preparations of exudate (e.g. sputum, pus) or biopsy specimens.
2. *Isolation* on appropriate culture media incubated at the correct temperature: some of the fungi grow slowly on culture.
3. *Serology:* useful in the diagnosis of histoplasmosis, coccidioido-mycosis and South American blastomycosis but of uncertain value in the other diseases because of difficulties in interpreting the significance of antibody levels.

Treatment

Amphotericin B is the drug of choice for invasive disease. Keto-conazole is the alternative treatment, at present under evaluation.

Medical parasitology

48

Parasitic infections

Parasites are much larger than bacteria. Some are multicellular, others single-cells, and they include the amoebae and worms with which most medical students are familiar. Man is not always their main host and may be merely an accidental host in their life cycle: some have intermediate hosts such as mosquitoes and snails in addition to the main definitive animal they inhabit. Clean water supplies and good food hygiene have reduced the risk of acquiring parasitic infection in Britain. On the other hand, the increase in foreign air travel and in the immigrant population have resulted in such infections being seen more commonly now than in the past.

PARASITIC INFECTIONS INDIGENOUS TO THE UNITED KINGDOM

Table 48.1 lists the main parasites that can be acquired in Britain.

TOXOCARIASIS

T.canis and *T.cati* are common in unwormed pets in Britain. Infection is transmitted from dog or cat faeces—especially to children. Usually asymptomatic but larvae may travel (visceral larva migrans) to eye (causing blindness) or liver, spleen or lungs with development of hepatosplenomegaly or pneumonitis. Eosinophilia and hepatomegaly are common.

Diagnosis

ELISA test (with secretory/excretory antigen).

Control

Worming of pets.

Table 48.1 Parasites indigenous to Britain

Parasite	Host	Source of Infection	Principal symptoms *
Nematodes (round worms)			
Toxocara canis	dog	dog & cat faeces	often asymptomatic, sometimes visceral larva migrans
Toxocara cati	cat		cerebral or ocular damage
Enterobius vermicularis	man	faecal-oral	asymptomatic, anal pruritus
Ascaris lumbricoides	man	faecal-oral	pneumonitis
Cestodes (tape worms)			
Echinococcus granulosus	dogs sheep	animal faeces	hydatid cysts especially liver and lung
Taenia saginata	cattle	raw, under-cooked meat	often asymptomatic, loss of weight, digestive disturbances
Trematodes (flukes)			
Fasciola hepatica	sheep	contaminated watercress	hepatitis, cholangitis, cholecystitis
Protozoa			
Giardia lamblia	man	contaminated water	diarrhoea
Trichomonas vaginalis	man	sexual transmission	vaginal discharge
Toxoplasma gondii	cat	cat faeces, raw, under-cooked meat	lymphadenopathy: congenital infection
Pneumocystis carinii	man	unknown	pneumonitis
Cryptosporidium	? farm animals, pets	faecal-oral	diarrhoea, abdominal pain, vomiting
Babesia divergens	cattle	tick-borne	fever, fatigue, haemolytic anaemia (especially in splenectomised patients)

* Note: symptomless infection is common with all parasites

Treatment

Thiabendazole, diethyl carbamazine.

THREAD OR PINWORMS

Not uncommon in children: infection with *Enterobius vermicularis* is usually symptomless but there may be anal itching with worms—recognised as 'white threads'—in stools.

Diagnosis

Demonstration of ova on perianal skin: worms or larvae in stools.

Treatment

Mebendazole, piperazine.

ROUND WORM

The nematode, *Ascaris lumbricoides,* inhabits the small intestine usually symptomlessly: if infestation is heavy, the sheer bulk of worms may cause symptoms of intestinal obstruction. There may be migration to lungs causing pneumonitis and to liver and other organs.

Diagnosis

Demonstration of ova in faeces.

Treatment

Mebendazole, levamisole, piperazine.

ECHINOCOCCUS GRANULOSIS

A rare infection in Britain, this cestode causes hydatid disease most often in the liver occasionally in lungs and other organs. Hydatid cysts are due to the migration of larvae which can be seen within well-encapsulated and sometimes very large cysts. Infection is acquired from dogs which carry the tapeworm: man becomes infected by ingestion of eggs.

Diagnosis

(i)   Serology—a variety of serological tests are in use including agglutination, complement fixation and ELISA tests.
(ii) Detection of protoscolices in cyst fluid.

Treatment

Surgical removal of cyst, mebendazole.

TAENIASIS

Now rare in Britain, infection with the tape-worm *Taenia saginata* is due to intestinal infestation with the adult worm. Segments are excreted in the faeces but symptoms are often mild—anorexia, loss of weight, abdominal pain, sometimes nervousness and insomnia. Infection is acquired by eating inadequately cooked beef muscle containing cysticerci: ova passed in the stool are infectious only for cattle.

Diagnosis

Detection of segments (proglottids) or ova in stools (or around the anus).

Treatment

Niclosamide.

FASCIOLIASIS

Fasciola hepatica is the liver fluke—a large parasite of which the natural host is sheep (or sometimes cattle). Infection in Britain is rare but cases have been traced to contaminated water-cress. The eggs in sheep faeces develop in water into motile larvae which infect snails—the intermediate hosts—and so aquatic plants like water-cress. Clinically the flukes live in the liver ducts and in the liver itself causing hepatitis, biliary colic and jaundice.

Diagnosis

Demonstration of eggs in stool, or in bile.

Treatment

Bithionol.

GIARDIASIS

An important cause of diarrhoea, the flagellated protozoan *Giardia lamblia* infects man through contamination of the water supply. There are two forms of the parasite—the trophozoite and cyst. Trophozoites are the motile feeding form and are responsible for symptoms in the host: found in the stool of an infected person because of rapid passage, but are not infectious and play no role in transmission. The cyst is the infectious form of the parasite. Water contaminated with cysts is a common source of infection but fomites contaminated with cysts may spread infection in schools, nurseries etc. Several water-borne outbreaks have been reported: Leningrad seems to be a particular focus of infection.

Diagnosis

Demonstration of cysts or trophozoites in stool.

Treatment

Quinacrine hydrochloride or metronidazole.

TRICHOMONIASIS

A major cause of vaginitis in women, with foul, yellowish, foamy vaginal discharge. *Trichomonas vaginalis* is a flagellated protozoan which is generally sexually transmitted from males with inapparent infection: contaminated articles may also be a source of infection.

Diagnosis

(i) Demonstration of motile parasites in vaginal secretions.
(ii) Culture: in Fineberg's medium.

Treatment

Metronidazole

TOXOPLASMOSIS

A systemic protozoal disease of which the most serious manifestation is congenital infection. The primary infection is a generalized disease, often asymptomatic but sometimes with fever and lymphadenopathy. Cysts of *Toxoplasma gondii* remain dormant in tissues—especially CNS and muscle—but including heart, and may reactivate later especially if immune deficiency supervenes. Primary infection in early pregnancy can lead to infection of the fetus or cause abortion. The infected baby may be born with chorioretinitis, cerebral calcification, microcephaly, and signs of generalised infection—fever, rash, jaundice and hepatosplenomegaly. If infection is later in pregnancy, the baby may escape infection altogether or be minimally infected but develop signs of chorioretinitis many years later. Toxoplasma is a parasite of all warm blooded animals—tissue cysts develop and human infection occurs when the flesh of such animals is eaten raw or very lightly cooked. The cat family is the definitive host of toxoplasma and in this species, sexual forms occur in the intestinal wall and eventually oocysts (which are also infective) are excreted. Contamination of sand pits etc. with cat faeces can lead to human infection.

Diagnosis

Primarily serological (e.g. ELISA for IgM, latex agglutination, dye, haemagglutination tests for IgG) but sometimes demonstration of the protozoan in tissues or fluids is possible.

Treatment

Pyrimethamine with sulphonamide and folinic acid.

PNEUMOCYSTIS CARINII

This protozoan infects a proportion of the normal population but without symptoms. The epidemiology is unknown but the parasite has been detected in many different species since it was first recognised in sewer rats. Infection can cause a severe pneumonitis in immunosuppressed patients and is a particular problem in cancer, transplant or AIDS patients.

Diagnosis

Demonstration in lung biopsy or bronchial washings of characteristic pneumocysts.

Serology

Detection of rising titres of antibody by immunofluorescence or ELISA may sometimes be of value.

Treatment

Cotrimoxazole, pentamidine

CRYPTOSPORIDIOSIS

Only recently discovered as a cause—apparently a common one—of diarrhoeal disease in children and adults. Adults are generally more severely affected than children with abdominal cramps a prominent feature. The epidemiology is unknown. There may be human case-to-case spread. Alternatively, infection may be acquired from domestic animals or pets.

Diagnosis

Demonstration of characteristic crytosporidium oocysts in stools by modified Ziehl-Neelson Method.

Treatment

No effective treatment is yet available.

BABESIOSIS

Babesiosis is rare but has been reported in the north of Scotland. *Babesia divergens* is a protozoan that infects cattle and human infection is transmitted through the bite of the tick *Ixodes ricinus*. It infects red blood cells and is particularly hazardous to the splenectomized in whom it can cause severe, even fatal haemolytic anaemia.

Diagnosis

Identification of parasite within erythrocytes in blood film.

Treatment

Of doubtful efficacy, pentamidine may be of benefit.

PARASITIC INFECTIONS NOT INDIGENOUS TO BRITAIN

Many of these are important and major causes of ill-health, particularly in the tropics and in underdeveloped countries. However, only malaria will be described in detail here. Malaria is not infrequently seen in Britain in travellers returned from abroad. It can easily be missed clinically—sometimes with serious results for the patient.

MALARIA

Malaria is no longer indigenous to Britain although it was once common in the Wash in the East of England. Transmitted by the bite of blood-sucking Anopheles mosquitoes, malaria is still a worldwide scourge and a major cause of ill health. Attempts to control it have been defeated by the development of drug resistance and by problems related to the organization of programmes for the control of mosquitoes in under-developed countries. Malaria is endemic in a huge belt across the world consisting of South America with the southern part of North America, Africa, the Middle and Far East and the islands of South East Asia.

The four main malaria parasites are shown below in Table 48.2

Table 48.2   The four main malaria parasites

Species	Distribution
Plasmodium vivax	the most common, found in all endemic areas and extending into sub-tropical and temperate zones
Plasmodium falciparum	also common, found in most endemic areas but not in temperate zones
Plasmodium malariae	much less common, mainly found in sub-tropical and temperate regions
Plasmodium ovale	predominant in West Africa, rare in other endemic areas

Malarial parasites have a complex life-cycle with different stages of development in man and mosquito. This is shown in a simplified form in Fig. 48.1.

Clinical Features

The main symptoms are of an intermittent fever which recurs at regular intervals depending on the parasite.

Periodicity

P.falciparum	36-48 hours	(malignant tertian)
P.vivax	48 hours	(benign tertian)
P.malariae	72 hours	(quartan)
P.ovale	48 hours	

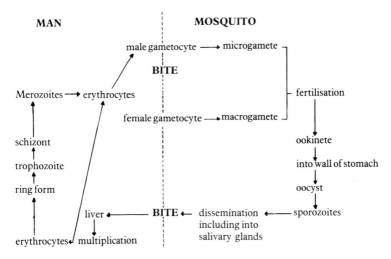

Fig.48.1 Malarial life-cycle in man and mosquito.

Incubation period

Variable from 8-40 days the longest being with *P.malariae,* the shortest, *P. falciparum.*

Prodrome

'Flu-like' symptoms e.g. headache, muscle pains, anorexia, photophobia, are sometimes seen at the end of the incubation period.

Malarial paroxysm

1. Coincides with lysis of infected erythrocytes, and liberation of merozoites.
2. Rigor or shaking chill, the patient complains of feeling cold but is in fact febrile.
3. Followed by feeling hot, flushed, agitated.
4. Severe headache and aching limbs and back are common.

Relapse

At the appropriate interval usually follows.

Pathogenesis

Symptoms are due to:

1. Haemolysis and the release of metabolites and pigment of malarial parasites.
2. Plugging of capillaries by parasites and infected erythrocytes.

Complications

1. *Cerebral malaria:* a complication mainly of *P.falciparum* infection, rarely, other forms of malaria.
   Symptoms are of headache, disorientation leading to coma and, if untreated, commonly to death.
   Cerebral infection due to spread of the parasites to the CNS.
2. *Blackwater fever:* also most often seen in *falciparum malaria.*
   Symptoms are of haemoglobinuria due to sudden intravascular haemolysis.
   Renal failure sometimes ensues due to acute tubular necrosis as a result of renal anoxia.
3. *Proteinuria:* is sometimes seen in infection with *P.malariae* when the kidneys of children are affected producing 'quartan nephrosis' with oedema of face and limbs. This serious complication of malaria infection has a prolonged course.
4. *Infection of placenta* in tropical Africa with *P. falciparum* leading to abortion, still-birth and low-weight babies.
5. *Tropical splenomegaly:* grossly enlarged spleen—an important complication of *P.malariae* infection.

Diagnosis

Demonstration of parasites in thick and thin blood films using Fields or Giemsa stain (thick film) and Giemsa stain (thin film).

Prophylaxis

Chloroquine, amodiaquine, primaquine, in areas with chloroquine-resistant *P.falciparum,* combined pyrimethamine and sulphadoxine.

Treatment

Chloroquine or amodiaquine. Alternatively quinone, pyrimethamine with sulphadoxine.

Other important parasitic infections found in countries other than Britain are listed in Table 48.2.

Table 48.3 Tropical parasites

Parasite	Host	Intermediate Host/Vector	Symptoms
Nematodes			
Hookworm- *Ancylostoma duodenale* *Necator americanus*	 Man	 — 	anaemia gastro-intestinal haemorrhage
Strongyloides stercoralis	Man	—	serpiginous skin lesions pneumonitis, enteritis
Trichuris trichiura	Man	—	diarrhoea
Trichinella spiralis	Pigs	Man	fever, muscle pain
Filaria			
Wuchereria bancrofti *Brugia malayi*	Man	Mosquito	lymphangitis elephantiasis
Onchocercus volvulus	Man	Blackfly	skin nodules ocular-blindness
Loa loa	Man	Fly	calabar swellings larvae in eye
Trematodes			
Schistosoma mansoni *japonicum* *haematobium*	 Man	 Snails	rectal bleeding, liver cirrhosis, portal hypertension haematuria
Paragonimus westermani	Cats Dogs	Shellfish	haemoptysis
Clonorchis sinensis	Cats etc.	Fish	cholangitis liver abscess
Protozoa			
Entamoeba histolytica	Man	—	amoebic dysentery with bloody diarrhoea, liver abscess
Trypanosoma rhodesiense *Trypanosoma gambiense*	Cattle Game animals, unknown	Tsetse fly	fever, meningoencephalitis
Trypanosoma cruzi	various domestic wild animals	bugs	Chagas' disease
Leishmania donovani *Leishmania tropica* *Leishmania braziliensis*	 Dogs Rodents	 Sand fly	cutaneous (oriental sore) visceral (kala azar)

RECOMMENDED READING

Benenson A S (ed) 1985 Control of communicable diseases in man 14th edn. American Public Health Association, Washington.

Christie A B 1981 Infectious diseases. 3rd edn. Churchill Livingstone, Edinburgh.

Wilson G S, Miles A, Parker M T 1984 Topley and Wilson's principles of bacteriology, virology and immunity, vols. 1, 2, 3 & 4. 7th edn. Edward Arnold, London

Vol. 1: General Microbiology and Immunology

Vol. 2: Systematic Bacteriology.

Vol. 3: Bacterial Diseases.

Vol. 4: Virology.

Index